THE MEANINGS OF A DISASTER

The Environment in History: International Perspectives

Series Editors: Stefania Barca, *University of Coimbra*; Christof Mauch, *LMU Munich*; Kieko Matteson, *University of Hawai'i at Mānoa*; Helmuth Trischler, *Deutsches Museum, Munich*

eseh european society for environmental history

Rachel Carson Center
ENVIRONMENT AND SOCIETY

Recent volumes:

Volume 20
The Meanings of a Disaster: Chernobyl and Its Afterlives in Britain and France
Karena Kalmbach

Volume 19
Conservation's Roots: Managing for Sustainability in Preindustrial Europe, 1100–1800
Edited by Abigail P. Dowling and Richard Keyser

Volume 18
Colonial Seeds in African Soil: A Critical History of Forest Conservation in Sierra Leone
Paul Munro

Volume 17
Hazardous Chemicals: Agents of Risk and Change, 1800–2000
Edited by Ernst Homburg and Elisabeth Vaupel

Volume 16
Planning for the Planet: Environmental Expertise and the International Union for Conservation of Nature and Natural Resources, 1960–1980
Simone Schleper

Volume 15
Changes in the Air: Hurricanes in New Orleans from 1718 to the Present
Eleonora Rohland

Volume 14
Ice and Snow in the Cold War: Histories of Extreme Climatic Environments
Edited by Julia Herzberg, Christian Kehrt, and Franziska Torma

Volume 13
A Living Past: Environmental Histories of Modern Latin America
Edited by John Soluri, Claudia Leal, and José Augusto Pádua

Volume 12
Managing Northern Europe's Forests: Histories from the Age of Improvement to the Age of Ecology
Edited by K. Jan Oosthoek and Richard Hölzl

Volume 11
International Organizations and Environmental Protection: Conservation and Globalization in the Twentieth Century
Edited by Wolfram Kaiser and Jan-Henrik Meyer

For a full volume listing, please see the series page on our website: http://berghahnbooks.com/series/environment-in-history

The Meanings of a Disaster

Chernobyl and Its Afterlives in Britain and France

Karena Kalmbach

berghahn
NEW YORK • OXFORD
www.berghahnbooks.com

First published in 2021 by
Berghahn Books
www.berghahnbooks.com

Parts of the text and arguments that appear in this book have been previously published in:

Bauer, S., Kalmbach, K., Kasperski, T. 'From Pripyat to Paris, from Grassroots Memories to Globalized Knowledge Production: The Politics of Chernobyl Fallout', in Laurel MacDowell (ed.), *Nuclear Portraits* (Toronto: University of Toronto Press, 2017), 149–89.

Kalmbach, Karena. *Tschernobyl und Frankreich: Die Debatte um die Auswirkungen des Reaktorunfalls im Kontext der französischen Atompolitik und Elitenkultur.* Frankfurt a. M., Peter Lang, 2011.

The author is grateful to Peter Lang International Academic Publishers and University of Toronto Press to grant the rights for this publication.

Library of Congress Cataloging-in-Publication Data

Names: Kalmbach, Karena, author.
Title: The meanings of a disaster : Chernobyl and its afterlives in Britain and France / Karena Kalmbach.
Description: New York : Berghahn Books, 2021. | Series: Environment in history: international perspectives; volume 20 | Includes bibliographical references and index.
Identifiers: LCCN 2020026211 (print) | LCCN 2020026212 (ebook) | ISBN 9781789207026 (hardback) | ISBN 9781789207033 (ebook)
Subjects: LCSH: Chernobyl Nuclear Accident, Chornobyl', Ukraine, 1986—Social aspects—Great Britain. | Chernobyl Nuclear Accident, Chornobyl', Ukraine, 1986—Social aspects—France. | Nuclear energy—Government policy—Great Britain. | Nuclear energy—Government policy—France.
Classification: LCC TK1362.U38 K365 2020 (print) | LCC TK1362.U38 (ebook) | DDC 363.17/99094777—dc23
LC record available at https://lccn.loc.gov/2020026211
LC ebook record available at https://lccn.loc.gov/2020026212

British Library Cataloguing in Publication Data

A catalogue record for this book is available from the British Library

ISBN 978-1-78920-702-6 hardback
ISBN 978-1-78920-703-3 ebook

Contents

Abbreviations

ACRO	Association pour le contrôle de la radioactivité dans l'Ouest
AEA	Atomic Energy Authority (short for: UKAEA)
AFMT	Association française des malades de la thyroïde
AGR	Advanced gas-cooled reactor
ASN	Autorité de sûreté nucléaire
BNES	British Nuclear Energy Society
BNFL	British Nuclear Fuels Limited
CCI	Chernobyl Children International
CCLL	Chernobyl Children's Lifeline
CCP (UK)	Chernobyl Children's Project UK
CEA	Commissariat à l'énergie atomique
CEGB	Central Electricity Generating Board
CERRIE	Committee Examining Radiation Risks from Internal Emitters
CND	Campaign for Nuclear Disarmament
COGEMA	Compagnie générale des matières nucléaires
CORE	Cooperation for Rehabilitation
CRIIRAD	Commission de recherche et d'information indépendantes sur la radioactivité
DGS	Direction générale de la santé
DGSNR	Direction générale de la sûreté nucléaire et de la radioprotection
DoE	Department of the Environment
DSIN	Direction de la sûreté des installations nucléaires
EC	European Communities
EDF	Electricité de France

END	European Nuclear Disarmament
EPR	European Pressurized Reactor
ETB	Enfants de Tchernobyl Bélarus
EU	European Union
FEFU	Fédération Echanges France Ukraine
FoE	Friends of the Earth
FRAMATOME	Franco-américaine de constructions atomiques
GSIEN	Groupement des scientifiques pour l'information sur l'énergie nucléaire
HPA	Health Protection Agency
HSMO	Her Majesty's Stationery Office
IAEA	International Atomic Energy Agency
IBB	Internationales Bildungs- und Begegnungswerk
ICRP	International Commission on Radiological Protection
InVS	Institut de veille sanitaire
IPSN	Institut de protection et de sûreté nucléaire
IRSN	Institut de radioprotection et de sûreté nucléaire
MAFF	Ministry of Agriculture, Fisheries and Food
MEP	Member of the European Parliament
NGO	Non-governmental Organization
NRPB	National Radiological Protection Board
OPRI	Office de protection contre les rayonnements ionisants
PWR	Pressurized Water Reactor
RBMK	Reaktor Bolshoy Moshchnosti Kanalnyy
SCPRI	Service central de protection contre les rayonnements ionisants
SCRAM	Scottish Campaign to Resist the Atomic Menace
STS	Science and Technology Studies
UKAEA	United Kingdom Atomic Energy Authority
UNSCEAR	United Nations Scientific Committee on the Effects of Atomic Radiation
WHO	World Health Organization

Introduction

After the 1986 nuclear disaster in Chernobyl, the name of this Ukrainian town became synonymous with *the worst accident ever to have occurred in the civil use of nuclear energy*. Chernobyl has retained this status ever since, that is, until 11 March 2011, when an earthquake and the resulting tsunami partially destroyed the Japanese nuclear power plant at Fukushima. The meltdown of the core at Chernobyl, classified as category 7 on the International Nuclear and Radiological Event Scale, was indeed considered the worst accident that could happen at a nuclear power plant. Technically, the actual meltdown is over. Chernobyl is seen as an event of the past, to which a start and an end were attributed by the technical evaluations following the evolution of the incident. However, its consequences are far from over. As with war, the scars of a nuclear catastrophe run deep; the aftermath is engraved in the environment, in people's bodies and in their memories. Signing a peace treaty does not bring an end to suffering; burying a destroyed reactor core under tons of concrete does not mean the evacuees can come home and simply forget what happened.

Comparing the Chernobyl disaster to a war scene is not just the result of a creative thinking process too strongly conditioned by my research. Many Ukrainian and Belarusian accounts narrate and interpret the struggle endured by firefighters and rescue workers as a battle against an enemy: the burning reactor. The victims, destruction and displacements provoked by this burning reactor have been linked to those caused by the Second World War. The asymmetry of such an equation may seem obvious when we recall the millions killed on the battlefields and murdered in the concentration camps, and this comparison might even seem inappropriate. But these narrations of Chernobyl do indeed exist, as does the metaphor of the *nuclear holocaust*. Chernobyl, however, is also described as a moderately serious industrial accident that caused a few deaths and slightly increased the probability that lethal cancers would occur in the exposed population, in other words a minor health impact compared to the annual number of deaths from road accidents or smoking cigarettes.

So, what does this phrase *worst accident ever to have occurred in the civil use of nuclear energy* actually mean? For some, Chernobyl is proof that this technology must be abandoned, sooner rather than later. Yet, for others, Chernobyl proves this technology is among the best that mankind has invented to date. How is it possible that the same event can be interpreted in such different ways? Precisely this question is the topic of this book. Some might argue: what is so surprising about the fact that a person in Belarus, having lost not only loved ones but also their home, and whose birthplace has been wiped off the map, would frame the event in a different light than a technocratic engineer in Vienna, tasked with calculating the probability that exactly the same accident will happen in a different nuclear power plant? It is hardly surprising at all that these two people give a different meaning to Chernobyl. Such an observation is scarcely enough to build a whole argument for a book. But what if we find these divergent interpretations in societies considered to be detached from the event, geographically as well as politically? What if we hear completely different narratives and interpretations of the causes and consequences of the accident, even among different groups within these societies? Would this constitute valid grounds for investigating the origin of these different narratives and interpretations, and for seeking explanations for how they came about and are constructed? I think so, which is why I aim to clarify the processes that led to these competing 'truths' circulating in public debates on Chernobyl.

My approach here to concepts such as narratives, interpretations and constructions is discourse analytical. I do not aim to add my own 'truth' to the many already circulating about Chernobyl. What is more, I am not in a position to judge which 'truth' is the most valid. Undoubtedly, many criteria could justify such a judgement: the scientific or political authority of the person or institution to deliver a given statement, or the number of people or institutions that quote this statement. But judging these competing 'truths' in such a manner would be like skimming the surface instead of investigating the discourse. Therefore, I do not ask: is this statement valid? but rather: why was a certain narrative disseminated at a certain time by a certain person or institution, and what is the meaning and significance of this narrative?

Why Compare France and Britain?

France and Britain[1] have a common history in that since the twentieth century, they have been the only nuclear powers in Western Europe. This factor, attributed the role of *tertium comparationis* in the comparison, not only shaped each country's technological history of the military use of the atom, but also strongly shaped both nations' civil nuclear engineering developments. As an offshoot of its military application, the civil use of nuclear energy was similarly governed by classified policies and surrounded in secrecy. Furthermore, nuclear power plants were never simply power plants, in France or in Britain. The inauguration of the first plants went hand in hand with intensive demonstrations of national pride; pride that they had been able to master this technology, that these genius engineers were citizens of their nation. The nuclear powers' shared status not only reflected the specific role attributed early on to the civil use of nuclear energy, it also meant they encountered criticism from sections of their respective societies questioning this technology. Although this critique differed between the two countries – in France it was directed at the civil use of nuclear energy, and in Britain the military use – important environmental and anti-nuclear movements sprang up in both, shaping their national discourse on nuclear technology. Public and political support for the civil use of nuclear energy has since fluctuated over time in both countries. Although their respective developments did not begin at the same time nor evolve identically, both countries have demonstrated strong political support for a *nuclear renaissance* since the early 2000s.

With so many similarities, we might expect their direct reactions to the 1986 disaster and the ensuing debates on its impact would be similar, too. Yet, this is only true for one aspect: the direct reactions in 1986, particularly from official sources like government or radiation protection agencies. However, the debates regarding the impact of Chernobyl that would unfold in the years to come could hardly have progressed more differently. Having been hit by comparable levels of radioactive fallout,[2] the people in both countries were assured there would be no health repercussions. Official statements released at the time specified: the accident had occurred too far away for there to be any effects, and even if a minimal quantity of airborne radionuclides had reached the countries, the associated risk levels would be marginal. Furthermore, according to these official statements, there were no grounds for questioning the safety of the national nuclear programme because the accident was due to a combination of very particular elements: the Soviet (hence inferior) technology, the faulty design of the reactor and the plant workers' human errors. In the following chapters, I will analyse these statements, how they were communicated and how the public, certain groups and individuals reacted to them. I should point out here, that despite France's and Britain's similar point of departure in 1986, their debates on the impact of Chernobyl developed in very different ways. In France, Chernobyl was assigned the role of a *lieu de mémoire*[3] (site of memory, which signifies a symbolic element of a community's memorial heritage) and became a common reference point in nuclear energy debates as well as wider public policy arguments. In Britain, on the other hand, Chernobyl was practically almost forgotten.

How is it possible that Chernobyl was allocated two diametrically opposed positions in French and British collective memory? What influential factors enabled the memory of Chernobyl to be kept alive in France, yet buried in Britain? Who were the stakeholders in this process? And what are the overarching frames of these developments, so essential for our understanding of the symbolic meanings of the two national Chernobyl debates? These questions and my research attempts to answer them form the core of this book. As this book ventures into generally un-researched terrain, sometimes I will have to raise new questions instead of providing answers. But I am sure that raising new questions will also contribute to a better understanding of the contested 'truths' surrounding Chernobyl.

Until recently, Chernobyl's impact on the collective memory of Western European societies was practically overlooked. Historians had held on to the idea that a history of Chernobyl only existed in those countries most (directly) affected by the radioactive fallout: Belarus, Ukraine and the Russian Federation. Some attention turned towards West Germany, where the anti-nuclear movement had just reached its peak shortly before Chernobyl; news of the accident inevitably evoked a very strong reaction there. However, even though several Western European governments and radiation protection agencies were quick to state that their countries had not been affected by any considerable radioactive fallout – whether true or not, does not concern us here – Western Europe experienced an intensive debate over Chernobyl's potential impact. Chernobyl has thus become a historical European event and we cannot equate its discursive impact just with its physical impact. But if it was not the amount of radioactive fallout that determined a specific country's views on Chernobyl's impact, what contextual factors shaped the national Chernobyl debates? I intend to answer precisely this question by comparing the French and British Chernobyl debates.

The Levels of Comparison

My comparative research occurs on several levels. Firstly, I compare the various narratives within each national[4] Chernobyl debate. Secondly, I compare the different debates in Britain and France. Finally, my comparison of these debates on several anniversaries of the accident reveals the changes and continuities over time.

To compare the narratives within national contexts, I focus on the categories of personal affectedness, radiophobia/apocalypse and anti-East European/anti-Soviet stereotypes. These categories are very much interdependent – a certain belief in one, directly influences statements on the others. As each category refers to a different issue addressed within the Chernobyl debate, I explain them separately. The three narrative elements I compare relate to: national nuclear politics (personal affectedness), general debates on the health impact of low-level radiation (radiophobia/apocalypse) and the Cold War setting (anti-East European/anti-Soviet stereotypes).

Firstly, the category of *personal affectedness* as used here, is the way actors consider themselves or their direct environment being impacted by the accident. This impact is either physical in terms of the radioactive fallout or has a less physical connotation in that the perceived impact can transfer the accident's scenario to national nuclear plants. The fear of eating contaminated vegetables bought at the local market or the fear that a similar accident could happen in a nearby nuclear plant, are the direct result of personal affectedness. Personal affectedness concerns the local, regional and national context. A strong perception of personal affectedness could cause an actor to call for a certain policy at national level, such as: banning certain foodstuffs, increasing control of radiation levels in the air and soil, instigating safety checks at national nuclear power plants, or even shutting down national plants. On the other hand, if an actor perceived low personal affectedness or even none at all, they considered these claims to be the product of panic or exaggerated fear and saw no reason why the accident in Chernobyl should influence national nuclear policies.

Secondly, the category of *radiophobia/apocalypse* tackles the different evaluations of the situation in Eastern Europe's regions most impacted by the radioactive fallout. The two extremes at either end of the evaluation scale are the explanatory concepts of radiophobia and apocalypse.[5] The radiophobia concept implies that the increase in illnesses observed in these regions is not actually down to the radiation itself; it is a result of the exaggerated fear of radiation and the psychological stress provoked by the resettlements and the rapidly changing political situation in Eastern Europe in the late 1980s and early 1990s. The concept implies furthermore that this stress was reinforced by distrust in the medical and political authorities, resulting in people's increased uncertainties and worries about their own health and that of their children. To avoid saying 'radiophobia' because this term was discredited in the early 1990s, sources often re-termed these assumptions as 'stress-induced illnesses'.[6] From a radiophobia point of view, the best cure for the illnesses would be to bring these regions back to normal and give the resettled populations incentives to move back to their home regions. The other end of the evaluation scale is an apocalyptic image.[7] This reading of the situation considers that the worst is yet to come: due to genetic mutations in humans, plants and animals, the true impact of the accident will only come to light very gradually, and there is nothing anyone can do to stop this process. From this perspective, measures must be taken to resettle even larger sections of the affected populations. These evaluations of the situation in Eastern Europe are, of course, closely linked to an actor's perception of personal affectedness: if people believe that exaggerated fear is the source of increasing illnesses in Belarus, they would not consider it even possible that the fallout in Western Europe had an impact on health there. At any rate, the frames of reference differ. Personal affectedness refers to the (Western European) actor's direct environment, whereas radiophobia and apocalypse focus on Eastern Europe.

Thirdly, the category of *anti-East European/anti-Soviet stereotypes* considers the way an actor draws on anti-East European or anti-Soviet stereotypes to describe the accident scenario. Using a kind of Cold War rhetoric, many narratives about Chernobyl comment on alcohol consumption and a generally imprudent handling of dangerous technologies within the USSR, contrasting them with statements on the good behaviour modelled in the West. Yet other actors outright dismiss these accusations, considering them as propaganda aimed to cover up problems in the West's nuclear initiatives. Thereby, Chernobyl narratives became a statement against the political opposition in the East–West discourse. Again, instead of judging which of these viewpoints is more or less 'true'[8] I here concentrate on how the legacy of Cold War propaganda and side-taking prevails in the Chernobyl debates and influenced interpretations of the accident. Compared to the two other categories of comparison, the aspect of anti-East European/anti-Soviet stereotypes plays a less prominent role in more recent sources. It featured primarily in accounts published in 1986 and 1987, addressing the acute phase of the accident: the event as it unfolded and the immediate aftermath. When my research focus switched from the technicalities to the debate surrounding the health impact of the fallout, the other two categories of comparison became more central to my analysis.

However, in March 2011, the issue of cultural stereotypes and reactions to nuclear plant accidents gained new momentum. Many actors whom I had researched regarding the Chernobyl debate initially narrated and framed Fukushima similarly to Chernobyl.[9] But one decisive element did not apply in 2011: the Cold War. This gap in the narrative structure, however, was quickly filled with something resembling the anti-East European/anti-Soviet element: the Japanese hierarchy bondage stereotype.[10] What now came under fire instead of 'Eastern European carefreeness', was the 'Japanese authoritative culture', which was held responsible for the course of events. This alternative explanation for the occurrence of a large-scale nuclear accident highly influenced the other discursive categories of personal affectedness and radiophobia/apocalypse in the Fukushima narratives. As stated above, a conviction regarding one variable directly influences the statements made about the others. In Western Europe, the perception of personal affectedness (in terms of how the accident's scenario is transferred to national nuclear plants) was even stronger in the case of Fukushima. The very fact that *stress tests* (safety assessments) were conducted on all the reactors in the European Union (EU), is telling evidence of this perception. The destroyed reactors in Chernobyl and Fukushima did indeed have very different technical features. Thus, technical considerations played a decisive role in implementing stress tests after Fukushima but not after Chernobyl. However, the fact that the Fukushima Daiichi reactor design was more like the Western European plants than the Chernobyl RBMK reactor, was not the main issue triggering Europeans' sudden safety concerns. The reasoning was rather: if such an accident can happen in a well-organized country like Japan, then a severe nuclear accident might very well be possible in the Western world too. With Chernobyl, the reasoning had been the opposite: since the accident happened in the USSR, it is impossible that a similar disaster could strike in the West. Contemplating this crucial difference in perceived Western European personal affectedness between Chernobyl and Fukushima directed my focus to the anti-East European/anti-Soviet stereotypes in the Chernobyl narratives. The events of 2011 have thus slightly changed one aspect of my research perspective but not my project as such. Although many people suggested I should switch to comparing the Chernobyl and Fukushima debates in one country rather than the Chernobyl debates in different countries, I persevered with my initial project. This decision was based on two factors: on the one hand, I think the Fukushima debate is still unfolding, which makes it difficult to achieve an accurate analysis of the narratives and interpretations. On the other hand, the constant and often not very well-informed references to Chernobyl within the Fukushima discourse have proved that the Chernobyl debate still crucially requires research.

Although I do not provide a general theory of how to analyse discourses on nuclear accidents, my analytic categories, findings and hypotheses can prove useful for researchers dealing with other (nuclear) disasters.[11] Both Fukushima and Chernobyl have demonstrated that, regardless of where an accident occurs, large-scale nuclear accidents are approached technically, socially and discursively in many parallel ways. However, my research is not just about Chernobyl and nuclear accidents. This book also sheds light on important aspects of the social and cultural history of France and Britain. The following section introduces the aspects I used as categories for comparing the two nations.

Why should we research national entities in relation to a nuclear power plant accident – an unsurpassed transnational event – whose fallout completely disregarded every border, even an Iron Curtain? Because it is the nation state's framework that mostly determines the debates on nuclear power in a society and consequently the related debates on Chernobyl. Transnational aspects clearly play an important role, especially regarding the exchange of information, expert evaluation, anti-nuclear activist networks, globally acting companies or lobby groups and international organizations.[12] However, how all these influences interact and impact a debate on a specific political topic such as nuclear energy, is very much dominated by legal and institutional aspects, and therefore by the nation state. The field of nuclear politics is a particularly good example. Although many aspects of this field are regulated at an international level, for example through the International Atomic Energy Agency (IAEA), the International Commission on Radiological Protection, or Euratom, the decision to build or close down nuclear plants is taken at national level. Neighbouring states might express concerns, like Austria in the debate over the Czech power plant Temelín. National authorities, however, make the ultimate decision – whether it is the German Bundesrat and Bundestag in the 2010–11 debate on *Laufzeitverlängerung* (extending the active lifespan of German nuclear power plants); the French President, Prime Minister and certain technocrats with the 1974 *Plan Messmer*; or the Austrian people, in the form of a referendum against activating the Zwentendorf power plant.

As stated above, comparing the national case studies of France and Britain is particularly fruitful thanks to their shared history as nuclear powers.[13] With so many similarities in their social and cultural nuclear past, it is even more interesting to question why the trajectory of the Chernobyl debates in these two countries has been so different; why did Chernobyl become a *lieu de mémoire* in France, but not in Britain? To answer this question, I compared the discursive frameworks surrounding these debates. I was interested to see whether different reference points shaped the various narratives in the two national contexts. Having identified the reference points and arguments in the debates, I researched their historical and political contexts to discover why a certain reference worked or was needed for narratives in one national context but not in another. I identified what I consider the key aspects that have influenced the different development of the Chernobyl debates in France and the UK: the formation, role and status of nuclear experts and 'counter experts'; the (changes to) national nuclear politics, policies and polities as well as their pro-nuclear or anti-nuclear orientations; the anti-nuclear movement's structure, political role and protest culture; the (issues with) national nuclear plants; and the importance of charitable organizations. I will return to all these aspects in due course, to analyse and contextualize the sources. I mention them here shortly to indicate my lines of comparison. For the historical and political context, I must introduce a term that features prominently in my arguments: *techno-political regime*. Gabrielle Hecht used this term in her book *The Radiance of France* to explain the different approaches pursued by the Commissariat à l'énergie atomique (CEA) and Electricité de France (EDF) to nuclear policies in the post-war period. My use of the term differs because I do not distinguish different regimes within one state. Nevertheless, I describe the same 'linked sets of individuals, practices, artefacts, programs, and ideologies'.[14] As *techno-political regimes* play an essential role in the historical and political context of the two countries, I use this term to cluster the above-mentioned aspects linked to a country's nuclear complex.

State of the Art in Chernobyl Research

The Chernobyl accident has only recently become a topic of historical research. Historians are therefore on the cusp of defining their role in Chernobyl studies. This role will not be limited to finding out what really happened on 26 April 1986, once the archives are opened. Historians can primarily contribute by adopting a critical approach to sources, that is to say questioning how a certain narrative developed and not just taking it for granted, or by investigating a narrative's impact on a society's collective memory. Historicizing Chernobyl does not mean this event must be banished to the past, denying that its radioactive, social and political fallout still has an impact today and might continue to do so for a long time. Historicizing Chernobyl not only implies that Chernobyl is considered an event whose meaning changes over time; it also asks why these changes occur, in which context, and who is instigating them.

The history and anthropology of Eastern Europe is surely a field that has addressed Chernobyl more than any other. In this field, Adriana Petryna's work has received the most attention internationally.[15] Recently, a research group based at the Centre for Contemporary History in Potsdam (Germany),[16] headed by Melanie Arndt, has made significant contributions.[17] There are too many important works in this field of research to mention here in detail.[18] Widening the close links between Chernobyl and Eastern European history is a very slow process. The 2016 international conference *Chernobyl – Turning Point or Catalyst? Changing Practices, Structures and Perceptions in Environmental Policy and Politics (1970s–1990s)*, organized by the Heinrich-Böll-Stiftung in Berlin, was an important step towards investigating Chernobyl's social and political impact across Europe.[19] Within wider Western European nuclear history, Chernobyl does not yet play a central role.[20] For my research on French and British Chernobyl debates, works by Gabrielle Hecht,[21] Sezin Topçu[22] and Brian Wynne[23] were therefore of utmost importance.

Other humanities and social sciences disciplines have of course researched Chernobyl. In particular, political scientists have from the outset investigated the direct impact of the accident on politics and policies. Angela Liberatore's work stands out as she did not limit her research to a single country, but compared Italy, Germany, France and even included the European Community and international organizations.[24] Sociologists have also focussed on Chernobyl, using the social dynamics surrounding the accident to underpin their theories with empirical evidence.[25] Chernobyl has been associated with Ulrich Beck's work in particular; not only because his famous book *Risk Society*[26] came out the month immediately after the accident, but also because in Beck's later article, directly connecting his theory to the event, he coined a term used ever since to denote Chernobyl: *anthropological shock*.[27] Sociologists have long been greatly interested in researching and theorising risk. Wolfgang Bonß identified the very foundation of risk sociology in the debates on nuclear technology.[28] Considered an incarnation of a risky technology, the nuclear complex has been an illustrative example of risk sociologists' arguments.[29] Social philosophers have also concerned themselves with the specific social implications of nuclear technology.[30] The public perception of nuclear risk is yet another field researched intensively by sociologists[31] and one that Christoph Hohenemser and Ortwin Renn have examined in the case of Chernobyl.[32] What has proved problematic is that some researchers take a very normative stance when investigating people's so-called overreactions or irrational behaviour and fail to reflect on the basis of their own judgements and the scientific facts they rely on.

Researchers in literature and media studies have also focussed on Chernobyl, investigating the visualization and artistic narration[33] for example in the successful video game *S.T.A.L.K.E.R.*[34] – the highly successful HBO miniseries *Chernobyl* will very likely give a boost to this research. In recent years, a growing body of academic work has emerged, looking at how people experience and visualize their visits to the forbidden zone. These visits are probably the result of marketing campaigns promoting the Chernobyl plant as a tourist attraction.[35] Since the 2011 nuclear accident in Fukushima,[36] scholars have been investigating Chernobyl in connection with other nuclear accidents,[37] and within the broader context of disaster studies.[38]

The research in the humanities and social sciences represents the minority of academic texts about Chernobyl. Numerous disciplines within the natural sciences and engineering have contributed more: nuclear physics, civil engineering, meteorology, geology, biology and nuclear medicine all focus on different aspects of this nuclear accident. Research projects range from analysing the physical reactions in the reactor and the movements of the airborne radionuclides, to the deposition of radionuclides in different geological settings and the uptake of radionuclides in plants and animals. Intensive research has of course studied the effects of Chernobyl radiation on the human body. Simply the list of journal articles featuring Chernobyl in the title would fill an entire book. International governmental and non-governmental organizations have condensed this multitude of Chernobyl studies into reports on the health and environmental effects and calculated the Chernobyl death toll.[39] These reports are crucial in international nuclear politics.[40] Assessing the health impact of Chernobyl automatically implies stating the health effects of reprocessing plants, nuclear power plants, uranium mines and of course, other nuclear accidents. Assuming a certain number of Chernobyl victims from the released levels of radionuclides in 1986 implies a certain number of Fukushima victims from the released levels of radionuclides in 2011.[41]

The Methodology

When I began studying Chernobyl in 2006, I was particularly interested in how its twentieth anniversary had been used to underpin political arguments regarding energy policies. Applying a 'politics of history'[42] approach, I considered what commemorative activities the various actors in the debate had carried out and thereby how the accident had been instrumentalized as a political argument. The actors discursively constructed a specific narrative of the historical event that corroborated their nuclear political claims. While tracing the French Chernobyl debate and its evolution from 1986 to 2006, I focussed on the accident's anniversary, as accounts of Chernobyl were not published on random dates. They mostly appeared on the date of the accident, 26 April. My search for primary sources began with the years 1986–87 (for immediate interpretations of the event's occurrence), and then the years 1996 and 2006; I was thus able to cover a broad time span and identify changes and continuities in the narratives. I applied the same research strategy to Britain.

In practice, adopting this research strategy meant I had to find all kinds of material containing a narrative of Chernobyl and aiming at a wide audience[43] in 1986–87 or on the accident's tenth or twentieth anniversaries. This approach had two major implications. Firstly, I did not consult any institutional archives. I was not interested in internal discussions within the governmental agencies or anti-nuclear groups involved; this would have been a different project. The main criterion for including a source in my research was its public availability.[44] Secondly, I researched a broad variety of actors communicating their Chernobyl narratives through an even broader variety of material: books, newspaper articles, leaflets, websites, films, documentaries, songs, photos, art exhibitions, speeches and more. Thus, my sources were often a combination of textual and visual elements: books and leaflets are illustrated with graphs, photographs, or artwork; a CD has an illustrated cover and an explanatory booklet. When analysing a source, I always considered these various elements and their interaction. My research does not include a comprehensive quantitative analysis of media content.[45] However, I did look into British and French news reporting to see what level of importance was attributed to Chernobyl compared to other topics, and which aspects were covered in 1986–87 as well as on the accident's anniversaries.

I applied the method of hermeneutical source analysis, involving three steps. First, I investigated a specific source's background: who was the author? When was the item published? Were the various authors linked at a personal or institutional level? The second step was analysing the narrative presented in the source. I queried: how are the causes and consequences of Chernobyl described? What kind of metaphors, references and explanatory frameworks do the authors use? What statements and wider interpretations are implied? After this analysis, I located each narrative within the wider context of the Chernobyl debate. I discovered that most narratives had an underlying structure consisting of three key elements that can be described as variables. Depending on an author's perspective of these three aspects, the Chernobyl narrative took on a certain form. I was thus able not only to compare the narratives, but also to map the discursive field in which the Chernobyl debate was embedded. Elements from other debates had entered this discursive field, making it possible to assign meaning to Chernobyl within a wider reference frame. The key elements I identified for any given Chernobyl narrative are: personal affectedness, radiophobia/apocalypse and anti-East European/anti-Soviet stereotypes. I introduced these three key elements earlier when discussing the levels of comparison. They clearly emerged during the research process and were not applied a priori to the sources.

Every account of Chernobyl provides a particular narrative of the accident and a certain interpretation of what happened, what is still happening and what will happen in the future. No matter how objective or subjective it claims to be, each account aims to make a certain statement about Chernobyl. I refer to the way these statements differ and contest as the *Chernobyl debate*, namely the variety of and relationship between statements, interpretations and narratives on Chernobyl that have circulated in public debates over time. What turns this variety into a debate is the setting; although every statement claims to represent the reality surrounding Chernobyl, they are conflicting because they are all based on different narrative elements that they claim to be facts.

This conceptualization of the Chernobyl debate is deeply influenced by Science and Technology Studies (STS). Within this field, categories such as scientific fact/truth or expert/counter expert/layperson are not considered fixed. This is why I often put these terms in quotation marks to make clear that I am not making normative statements. Rather, I see these concepts and terms as part of a societal discourse and specific power setting. This, however, does not mean that I regard every statement ever pronounced on Chernobyl as equal to all the others or declare that one is just as valid as any other. A reviewer of my earlier book on Chernobyl and France claimed that I placed 'obscure theories about the course of the accident on a par with the relevant literature'.[46] According to the nuclear engineering scientific community's internal logic, it is certainly incorrect behaviour to dedicate equal space to both. However, my argument was not and still is not about judging these statements, my argument is about questioning the logic behind them and investigating the reasons for which they are presented.

Asking why these narratives were presented in a specific geographic location, at a specific time, brings us conveniently to the final step of the research process: the contextualization. Contextualization implies researching to what extent the French and British cultural-historical settings and political systems affected the development of the Chernobyl debates in each country. It also implies investigating how the national Chernobyl debates interconnect with international and transnational debates regarding the civil use of nuclear energy. Yet, what I do not provide in this book is an investigation into the history of my very categories of comparison. This could be considered a limitation of my work. But as this book is primarily concerned with nuclear debates in France and Britain, it would have exceeded the scope of this text to also elaborate the historiography of medical and psychological research on radiophobia or the technical grounds for anti-Soviet stereotypes in nuclear industry debates.[47] Similarly, the book does not cover the global history of the medical discourse on low-level radiation health effects or the history of radiation protection – which form the background of the debates on the health impact of Chernobyl.[48] It is the public French and British Chernobyl discourse which lies at the heart of this book, and I invite every reader to dive deeper into the various histories to which my particular Chernobyl story is connected.

This methodological approach shapes the structure of my book: I focus on the period between 1986, the year of the accident and 2006, its twentieth anniversary. I do not include the twenty-fifth anniversary in 2011, because the memory work on Chernobyl as well as the entire nuclear discourse were then strongly influenced and overshadowed by the recent Fukushima accident.[49] With regard to the actors, I have assembled them in 'actor clusters'. This was not only to avoid confusion and reduce the risk that the reader would get lost in an endless list of individuals and organizations; it was also a good way to highlight the similar backgrounds to the narratives emerging from these clusters. Based on the structural reasons for an actor's involvement in the Chernobyl debate, I identified the following clusters: public authorities (governments, radiation protection agencies); the nuclear power industry (companies, associations), critical voices (anti-nuclear groups, sheep farmers, landowners); individual voices; and Chernobyl solidarity movement groups. The key narrative categories of personal affectedness, radiophobia/apocalypse and anti-East European/anti-Soviet stereotypes are the other themes in this book. Throughout the text, I interweave my explanations and hypotheses on why different narratives have developed in French and British Chernobyl debates. Each chapter highlights the context in which a certain actor cluster operates. This context refers to the formation, role and status of nuclear experts and 'counter experts'; the (changes to) national nuclear politics, policies and polities as well as their pro-nuclear or anti-nuclear orientations; the anti-nuclear movement's structure, political role and protest culture; the (issues with) national nuclear plants; and the importance of charitable organizations. My explanations of this context are combined throughout the book with an analysis of the sources.

I decided to devote more space to British sources than French ones, for two reasons: firstly, my book on the French Chernobyl debate already deals with the French material in detail, and secondly I considered it important to discuss the British Chernobyl debate in depth and introduce as many actors as possible since this has not yet been a topic of historical research. This is clearly evidenced by the fact that a search for the key word 'Chernobyl' in the Bibliography of British and Irish History in 2012, when I started my research on Britain, showed zero results. In addition, the civil use of nuclear energy in Britain has been underrepresented in nuclear history, compared to works on military use.[50] The two main British nuclear historians, Margaret Gowing and Lorna Arnold,[51] have not covered the recent era and only a handful of publications analyse the British nuclear complex through a historical lens.[52] From the humanities and social science perspectives, only Brian Wynne has written about Britain and Chernobyl, albeit with a clear STS perspective on knowledge production and experts-laypersons-relations regarding the early restrictions on sheep farms. In order to pay due credit to the wider context of the French and British debates, I have incorporated in the text information on international and transnational aspects of the Chernobyl debate – such as the influence of dissident voices from Eastern Europe, the contestation of reports such as the one delivered by the Chernobyl Forum or the theories surrounding the so-called IAEA–WHO agreement.[53]

Notes

1. I use 'Britain' here to refer to The United Kingdom of Great Britain and Northern Ireland, or in short United Kingdom, or the UK.
2. For an early estimate of the levels and intensity of the radioactive fallout in both countries, see: NRPB, *Preliminary Assessment of the Radiological Impact*, Appendix, Figure 7. According to this map, French and British citizens received a similar effective dose: 10–50 μSv in most areas and 80–300 μSv in the more affected regions, specifically the southeast (France) and the north-west (Britain). Initial maps were produced by extrapolating individual measuring points (n.b. the number of points varied widely between countries). These estimates were revised and refined after 1986. Later maps were the result of a compendium of data derived for example from measuring radionuclides in soil and grass samples, rainfall or even nationwide caesium levels. In 1996, the European Commission published the comprehensive *Atlas on the Caesium Deposition across Europe*.
3. Since Pierre Nora's classic work on French sites of memory, the application of his concept shifted from nation building processes to widely ranging settings where collective memory contributes to a specific group's self-conception. My description of Chernobyl as a *lieu de mémoire* does not imply that Chernobyl holds a certain place in a nation's self-conception (though this might be the case, for example in Belarus) but rather that the memory of the event contributes to the identity of a certain group. Recently, the *lieu de mémoire* concept was adopted by environmental historians, resulting in an anthol-

ogy and a special issue of *Global Environment* on *ecological lieux de mémoire*. I contributed to both: Kalmbach, 'Radiation and Borders'; id., 'Von Strahlen und Grenzen'. I also contributed to the anthology on *European lieux de mémoire*: Kalmbach, 'Tchernobyl – angle mort'.

4. The phrase 'national Chernobyl debate' does not imply that debates necessarily took place on a national level. Many addressed regional issues like the environmental impact in the Lake District (UK) or in Corsica (France). Some issues remained there, while others attracted wider public interest. Thus, 'national Chernobyl debate' refers to the larger contextual setting and does not judge the scale of the debate.
5. If I classify a narrative as linked to the concept of 'radiophobia', I do not automatically imply that its author advocates that radiophobia exists as a medical condition. Likewise, by classifying a narrative as 'apocalyptic', I do not mean to imply that this narrative includes mutated monsters. I use the classifications 'apocalypse' and 'radiophobia' as an *Idealtypus*.
6. As Tatiana Kasperski demonstrated in her research, the term 'radiophobia' in relation to Chernobyl was first used in a report by Ilin, president of the USSR's national radiation protection committee, and his colleague Pavlovskii. This report was published in the 1987 4th IAEA Bulletin under the title *Radiological Consequences of the Chernobyl Accident in the Soviet Union and Measures Taken to Mitigate Their Impact: Analysis of Data Confirms the Effectiveness of Large-Scale Actions to Limit the Accident's Effects*. In Belarus, the term was profoundly discredited in the 1990s, as Kasperski explains in 'La politique de la mémoire': 'The protest movements which emerged in the late 80s and early 90s and for which the Chernobyl catastrophe served as catalyst often referred to this term in order to demonstrate the Soviet authorities' cynicism vis-à-vis the dangers the inhabitants and the liquidators had encountered. The term "radiophobia" thus became widely discredited because it evoked the attempts of the communist leaders to dissimulate the harm caused by the accident by misinforming the population and the international community' (393).
7. Narrating Chernobyl as an apocalypse must be placed in the wider context of traditionally framing disasters and catastrophes with apocalyptic images and of the deep impact of depictions of the destruction of Hiroshima and Nagasaki on the imagery surrounding nuclear technology in general. For a history of nuclear images, see: Weart, *Nuclear Fear*.
8. For a well-informed account of the security culture at the Chernobyl plant, see: Schmid, 'When Safe Enough is not Good Enough'. Schmid's work also shows that the use of national and cultural stereotypes in Chernobyl narratives had yet another dimension in the Soviet Union: the government in Moscow placed the blame on Ukrainian plant workers. However, the Chernobyl debates in Western Europe featured hardly any differentiation between groups and ethnicities in Eastern Europe or within the Soviet Union. For this reason, I did not consider the differentiation in my comparison.
9. Kalmbach, 'From Chernobyl to Fukushima'.
10. The interpretation of Fukushima as a 'Japanese disaster' was certified by the *Nuclear Accident Independent Investigation Commission* executive summary report in English. In his foreword, the chairman of this Diet-appointed commission, Kiyoshi Kurokawa, called the events at Fukushima a 'disaster "Made in Japan"': The National Diet of Japan, *The Official Report*, 9.

11. This book contributes to disaster research by highlighting the enduring aftermath of Chernobyl, and the important part narratives and memory play in a society's understanding of a disaster and its meaning for that society. As a disaster does not correspond with the precise moment the emergency measures are terminated, a political process is needed to negotiate a new 'normality' of daily life that can compensate for the effects of the actual disaster.
12. For a recent publication stressing the importance of transnational approaches in nuclear history, see: Kirchhof and Meyer, 'Global Protest against Nuclear Power'.
13. In considering a *tertium comparationis* other than nuclear power status, it could be worthwhile conducting comparative research on Chernobyl debates in Sweden and Turkey, two non-USSR countries that experienced significant levels of fallout, or in Denmark and Germany, two countries where anti-nuclear convictions were particularly strong in 1986.
14. Hecht, *The Radiance of France*, 56.
15. Petryna, *Life Exposed.*
16. The project group of six scholars conducted research on *Politics and Society after Chernobyl* in Belarus, Ukraine, Russia, Lithuania and Germany: Melanie Arndt, Evgenija Ivanova, Tatjana Kasperski, Anastasija Leuchina, Andrej Stepanov and Aleksandr Dalhouski.
17. Arndt, *Politik und Gesellschaft nach Tschernobyl.*
18. For the history and anthropology of Eastern Europe relating to Chernobyl, see: Arndt, 'Memories, Commemorations, and Representations of Chernobyl'; Kuchinskaya, 'Twice Invisible'; id., *The Politics of Invisibility*; Kasperski, 'La politique de la mémoire'; Stsiapanau, 'The Chernobyl Politics in Belarus'; Phillips, 'Chernobyl's Sixth Sense'; Sahm, *Transformation im Schatten von Tschernobyl*; id., 'Und der dritte Weltkrieg heißt Tschernobyl'; Arndt, 'Von der Todeszone zum Strahlen-Mekka'; Sahm and Sapper and Weichsel, 'Tschernobyl: Vermächtnis und Verpflichtung'; Marples, *Chernobyl and Nuclear Power*; id., *The Social Impact*; Schmid, *Producing Power*.
19. The presentations were filmed and are available online: https://www.youtube.com/watch?v=KpW5n9GVOtg (last accessed 15 February 2020).
20. Kalmbach, 'Revisiting the Nuclear Age'. For a media history study of Chernobyl, see: Jordan, *Ausgestrahlt.*
21. Hecht, *The Radiance of France*; id., 'Nuclear Ontologies'; id., *Being Nuclear.*
22. Topçu, *La France nucléaire*; id., *L'agir contestataire.*
23. Wynne, 'Misunderstood Misunderstanding'; id., 'Sheepfarming after Chernobyl'; id., *Rationality and Ritual.*
24. Liberatore, *The Management of Uncertainty.*
25. A telling example: Bonß, *Vom Risiko.*
26. Beck, *Risikogesellschaft.*
27. Beck, 'The Anthropological Shock'.
28. Bonß, *Vom Risiko*, 9.
29. See: Perrow, *Normal Accidents*; Luhmann, *Risk*; Freudenberg, 'Perceived Risk, Real Risk'; Slovic, 'Perception of Risk'; Boudia and Jas, 'Risk and "Risk Society"'.
30. See: Anders, *Endzeit und Zeitenende*; id., *Die atomare Drohung*; Dupuy, *Pour un catastrophisme éclairé.*

31. For a classic work, see: van der Pligt, *Nuclear Energy and the Public*. For a study dealing specifically with France, see: Bouvier, 'Risques perçus et risques industriels'. For how this perception is influenced by risk communication, see: Fischhoff, 'The Nuclear Energy Industry's Communication Problem'; id., 'Risk Perception and Communication Unplugged'.
32. Hohenemser and Renn, 'Chernobyl's Other Legacy'; Renn, 'Public Responses to the Chernobyl Accident'.
33. See: Bürkner, 'Eine vollkommen neue Realität'; Zink, 'Approaching the Void'.
34. *S.T.A.L.K.E.R.: Shadow of Chernobyl* is a popular ego-shooter video game set in the restricted zone after a second severe accident at Chernobyl. It was created by a team of Ukrainian game developers and released in 2007. Thanks to widespread success in Eastern and Western Europe and the US, a sequel was released in 2010: *S.T.A.L.K.E.R.: Call of Pripyat*.
35. For an analysis of tourists' photos of Chernobyl, see: Goatcher and Brunsden, 'Chernobyl and the Sublime Tourist'. For reflections on the role of Chernobyl as tourist attraction, see: Stone, 'Dark Tourism'.
36. Hindmarsh, *Nuclear Disaster at Fukushima Daiichi*.
37. See: Bohn et al., *The Impact of Disaster*.
38. See the 2012 workshop *Historical and Contemporary Studies of Disasters: Placing Chernobyl, 9/11, Katrina, Deepwater Horizon, Fukushima and Other Events in Historical and Comparative Perspective* (Society for the History of Technology annual conference) at: http://shotprometheans.wordpress.com/workshops/2012-workshop/ (last accessed 15 February 2020). In recent years, Science and Technology Studies scholars have made considerable efforts to establish a transnational research agenda for the new field of disaster STS: http://fukushimaforum.wordpress.com/workshops/sts-forum-on-the-2011-fukushima-east-japan-disaster/ (last accessed 15 February 2020).
39. The major reports by international (governmental and non-governmental) organizations on the health impact of Chernobyl are: WHO, *Health Effects of the Chernobyl Accident*; UNSCEAR, *Sources and Effects of Ionizing Radiation*; UNDP and UNICEF, *The Human Consequences*; The International Chernobyl Project, *Assessment of Radiological Consequences*; The Chernobyl Forum, *Chernobyl's Legacy*; Fairlie and Sumner, *The Other Report on Chernobyl*; Greenpeace, *The Chernobyl Catastrophe*; German Affiliate of International Physicians for the Prevention of Nuclear War (IPPNW) and Gesellschaft für Strahlenschutz, *Health Effects of Chernobyl*.
40. Bauer and Kalmbach and Kasperki, 'From Pripyat to Paris'; Brown, *Manual for Survival*.
41. The film *Fukushima and Chernobyl: Myth versus Reality* is a telling example. This film was commissioned by the London-based *World Nuclear Association*, until 2001 the *Uranium Institute*. It illustrates the tremendous importance of the debate on low-level radiation health effects for the pro-nuclear side after 2011: promoting radiophobia narratives on Chernobyl was a type of armour to deflect anti-nuclear voices, who, alarmed by Fukushima, called for a reassessment of the global nuclear enterprise. For an analysis of this film, see: Kalmbach, 'The Contested Truth', 271–9.
42. My use of the phrase 'politics of history' implies the discursive usage of historical events, persons and so forth as political arguments.
43. Articles exclusively published in scientific journals are thus not included in the sources considered here.

44. Thanks to the legal deposit ruling in both France and Britain, I could directly access material in the Bibliothèque nationale de France (BNF) in Paris and the British Library (BL) in London. Both libraries have extensive collections of grey literature, government publications and audio-visual material. Moreover, the BNF (site François Mitterrand) incorporates the Institut national de l'audiovisuel (INA), where I was able to research French TV-reports on Chernobyl. With regard to newspaper reporting, I researched the various databases at the BL, and Science Po Paris' press clippings. Fondation EDF, Chernobyl solidarity groups and individual activists kindly provided me with published grey literature not available in libraries.
45. I do not consider the media as actors. I prefer to study the actions of individual journalists and how actors' statements are reflected in media reporting. The media can be actors in public debates, particularly for agenda setting. But this aspect is less relevant as I focus on the events of 1986 and anniversaries.
46. Wendland, 'Rezension zu: Kalmbach, Karena'.
47. See for these perspectives the work of Anna Veronika Wendland, for instance: Wendland, 'Reaktorsicherheit als Zukunftskommunikation'.
48. See for instance: Onaga, 'Reconstructing the Linear No-Threshold Model in Japan'.
49. For an account of the 2011 transnational Chernobyl debate, see: Kalmbach, 'The Contested Truth', 267–88.
50. For important works on Britain's military nuclear history, see: Twigge and Scott, *Planning Armageddon*; Heuser, *Nuclear Mentalities*.
51. Gowing, *Independence and Deterrence*; id., *Britain and Atomic Energy*.
52. The most recent works on British nuclear history are: Laucht, *Elemental Germans*; Hogg and Laucht, 'British Nuclear Culture'. Ian Welsh's work on British anti-nuclear movements applies a social science perspective: Welsh, *Mobilising Modernity*; id., 'The NIMBY Syndrome'.
53. Concerning actors, I use the term *international* to refer to international governmental organizations (IGOs) and *transnational* for the activities of non-governmental organizations (NGOs) and civil society initiatives that transcend national borders.

CHAPTER 1

1986–1988

Direct Reactions and Early Narratives

On the night of 25–6 April 1986, the Lenin nuclear power plant, situated approximately 100 km north of the Ukrainian capital Kiev, was the scene of an accident that would go down in history under the name of the nearby town Chernobyl. The Lenin plant consisted of four RBMK model[1] reactor units, each capable of generating up to 1,000 Megawatts. The plant had been built in the 1970s and 1980s, and two new blocks were under construction in 1986. The nearby city of Pripyat was founded at the same time to serve the power plant. One of the most modern cities at that time in the USSR, Pripyat attracted young engineers and their families from all over the country. This symbiotic industrial-living-complex was considered the materialization of the USSR's technical progress. On this night in April, during a systems test, a series of explosions occurred in unit number four. The cooling system broke down and the graphite, used as a moderator in RBMK reactors, caught fire. As the thermal explosions destroyed the building housing the reactor, radioactive material was released directly into the environment.[2] The fires generated smoke and dust that carried radioactive particles high into the air, leading to a global dispersion of the radionuclides. The intensity of local contamination, however, depended on more than just the movement of the air masses. It also depended on the meteorological constellations and geographical settings. Chernobyl was not only a transnational event in regard its physical fallout. The media coverage also spread on a global scale. The USSR public only read about the events on 28 April, in a news brief released by the state press agency *TASS*, and no further information was provided. Several days later, Soviet television reported two deaths and declared that the radiation situation had been stabilized. At the same time, the media in the West, alarmed by the detection of radioactive fallout in Scandinavia, were already speculating about the causes of the accident and the possible total death toll. While the emergency measures were still ongoing, anti- and pro-nuclear activists outside the USSR were already fighting over the interpretation and political consequences of Chernobyl. Firefighters were only able to extinguish the graphite fires days later, after which clean-up workers moved in to replace them. These 'liquidators', as they were called, were responsible for decontaminating the other plant buildings, but also erected the 'sarcophagus' – the containment building that even today encloses destroyed reactor number four – and buried the contaminated soil, machines and debris. In all, more than 600,000 men and women were called to Chernobyl to work in the cleanup crews. Although the other three units were reconnected to the grid by the end of the year, an area with a radius of 30 km around the plant was declared a 'forbidden zone' and placed under military control. All the people in this region were evacuated, most of whom hoped and believed that they would soon return to their homes. But that would never happen. Pripyat became a ghost town and the smaller settlements in this region were demolished in the ensuing years. As measurements of the radioactivity in the environment continued to reveal dangerously high levels in the most contaminated areas of Belarus – the country hit with the most intense fallout – the initial number of 116,000 evacuees climbed to 350,000.[3]

Initial Media Reporting

The news of the accident in Chernobyl reached France and the UK at the same time. In France, the *Première cohabitation* – with François Mitterrand as President and Jacques Chirac as Prime Minister – was confronted with the events in the Ukraine, while in the UK, Margaret Thatcher's government received the news. Both nations' public authorities initially proclaimed the radioactive fallout would not have any serious impact on their country. This statement was profoundly called into question in France, resulting in an intense debate on whether the authorities had deliberately held back the true figures of the fallout intensity, a polemic referred to as the *affaire Tchernobyl*. In Britain, strong restrictions had to be implemented on sheep farms some weeks after the accident. Apparently, highland sheep had become too radioactive to be marketed after grazing on contaminated soil; however, these restrictions did not cause a long-lasting public scandal. This chapter focuses on why the dangers associated with the radioactive fallout deposited in France and Britain as well as the public experts' role in evaluating this fallout were perceived so differently in these countries. The French and British news reports discussed below show how the public first learned about the accident.

In order to dive into the situation in Britain in 1986, this section provides a selection of the early British news reports on Chernobyl, published in the daily national broadsheet newspaper the *Guardian*.[4] 'Radioactive Russian Dust Cloud Escapes' was the headline presented to the British public on 29 April; it was the first article the *Guardian* released about the Chernobyl disaster. Readers were informed that 'a major nuclear power accident in the Soviet Union yesterday sent a cloud of radioactivity drifting across much of Scandinavia'.[5] The article went on to quote the *TASS* announcement: 'An accident has occurred at the Chernobyl nuclear power plant as one of the reactors was damaged. Measures are being taken to eliminate the consequences of the accident. Aid is being given to those affected. A government commission has been set up'.[6]

Apart from reporting on the radiation levels in Scandinavia, the article focussed on the technical details regarding radioactive fallout detection, not the potential health hazards. Hence, a positive story of British preparedness was provided: 'If the Soviet plume begins to drift towards Britain – and there have been easterly winds – the National Radiological Protection will quickly pick up the signs from its fall-out monitoring stations'.[7]

The articles in the *Guardian* the following day addressed other aspects of the accident. Not only was the situation at the site of the accident itself discussed,[8] but also the context of global energy policies,[9] and the potential impact of this disaster on national movements in Belarus, Ukraine, Latvia and Lithuania.[10] An entire article was dedicated to the particularities of the RBMK reactor, interestingly mentioning that 'it was also a graphite-uranium core which caught fire in the Windscale accident of 1957, releasing large amounts of radioactive iodine and other materials over a wide area'.[11] Comparisons to the Windscale Fire – the worst accident ever to have occurred at a British nuclear plant[12] – were also made in this article to describe the ongoing situation at Chernobyl, touching on such aspects as the levels of radioactivity, the size of particles released into the environment and the actions taken to extinguish the fires. The author of this article was rather critical of the first official statements, asserting that: 'although it is being claimed by Western "experts" that all power reactors in the West have secondary containment, this is not true. Britain's Magnox reactors – which have dominated the nuclear programme – are without secondary containment because it was regarded as unnecessary at the time they were designed'.[13]

In another article, journalist David Fairhall took his analysis of the official statements on Chernobyl a step further, questioning the reasons for the strategy to distantiate the RBMK reactor and events in the Soviet Union from the actual situation in the UK:

> The winds in Russia are blowing eastwards, not westwards towards the UK, and the burning Chernobyl reactor is of a type not used in the West. So can we now relax and get on with our own nuclear power programme without worrying about the Soviet disaster? That is certainly what the British nuclear industry will be recommending. They will point out that the Chernobyl plant is a peculiarly Soviet design ...[14]

Thus, in the immediate aftermath of the accident at Chernobyl, the articles in the *Guardian* already made a direct connection between Chernobyl and the

British nuclear enterprise. The event was first explained by comparison to the national Windscale Fire incident, then the articles enquired how the Chernobyl accident would impact the new build Pressurized Water Reactor (PWR) project in Sizewell. With regard to national nuclear policies, the journalists expressed a strong perception of personal affectedness. With regard to the possible (health) effects of the fallout in Western Europe, however, their perceived personal affectedness was quite low: 'In Britain it seems unlikely that we shall feel any effects at all'.[15]

Reporting over the next days was primarily concerned with the British students evacuated from Kiev,[16] speculations on the death toll[17] and the debate over whether the accident should impact the new build project Sizewell B,[18] an openly contentious issue within the Labour Party.[19] Although the potential impact on health in the UK was further discussed, it did not dominate the headlines. Regardless of the fact that medical correspondent Andrew Vletch's article 'Parents Scour Chemists Shops for Iodine Tablets' clearly stated that there would be health effects in the UK, albeit very minor ones, his column only made page six on 2 May. According to Vletch, 'all it [the low-dose radiation from Chernobyl] will do is increase the incidence of cancer by an undetectable amount over the next 20 or 30 years'.[20] Interestingly, however, this article quoted Joseph Rotblat's statement, 'it would be nonsense to start taking iodine tablets. People are panicking because the reports of what has happened in Russia have been exaggerated'. This all-clear came from the most well-known opponent of nuclear weapons, a man who had received the Nobel Peace Prize for his efforts to achieve a nuclear weapons test ban and nuclear disarmament, and who had worked intensively on the health impact of nuclear fallout. Without a doubt, his statement was received within the circle of nuclear critics as utmost credible. The front-page article the very next day in the *Guardian* (3 May) corroborated this rather untroubled stance on radioactivity levels in Britain. The article, headlined 'Radiation Cloud Reaches Britain', informed readers that 'the radioactive cloud from the Soviet nuclear disaster reached Britain yesterday. But the National Radiological Protection Board [NRPB] described the contamination level as very low and the Department of Health said it posed no health risk to the public'.[21] The article quoted Donald Acheson, the Department of Health's Chief Medical Officer, who stated: 'there was absolutely no need to take potassium iodine tablets' and 'it was most unlikely that those returning home from affected areas were at risk'.[22] Similar reporting continued on into the next day: 'Radioactivity from the Chernobyl disaster was still being recorded all over Britain yesterday. Nationwide checks are being made on milk after traces were found in supplies. The National Radiological Protection Board emphasized that there was no danger'.[23]

After heavy rainfall and thunderstorms in some parts of the UK on 3, 4 and 5 May had 'washed' the radionuclides to the ground, the NRPB slightly changed its evaluation and began advising that 'rainwater should not be drunk if it can be avoided'.[24] However, this advice was not considered front-page news by the *Guardian* and was stuck half a page down an article comparing countermeasures taken all over Europe. Eventually, on 7 May, the issue of fallout personal affectedness gained momentum, as did information policies: 'Radioactivity Alarms Scotch Makers',[25] 'Baker Acts to Soothe Radioactivity Fears',[26] 'MPs Told: There is No Health Risk',[27] 'Levels Dropping Fast as Cloud Moves away',[28] 'Contaminated Rain Sets Welsh Alarms Ringing',[29] read the headlines. To deal with rising concerns and calm people's anxiety, a special incident room was set up which the public could call for advice on foodstuff. The Chernobyl accident slowly transformed into a British political issue, not only because the opposition protested 'that there had been considerable confusion about the advice being offered by officials'[30] but also because opinion polls showed a rapid increase in people's opposition to the government's nuclear new build projects.[31] These polls also revealed a 'crisis of credibility': 'At least half the population, if the polls are right, does not accept that the new radioactivity is harmless'.[32] Hugo Young, in his comment on 8 May, partly attributed this lack of trust[33] to the nuclear 'tradition from which his [Mr Baker, the Environment Secretary] own honesty cannot disconnect … the folk-memory is filled more with lies than with truth: lies about the Windscale fire, evasions about numerous subsequent incidents, official documents spelling out a calculated policy of misinformation and subterfuge.'[34]

Young believed this mistrust went far beyond the specific case of Chernobyl and needed to be explained in light of general problematic issues in the British political system:

> If ministers are not trusted, it is because they personify a world which has always taken a minimalist approach to public information. This world, of Whitehall and Higher Westminster, sees the media as organs to be feared or manipulated but not respected: sees politicians themselves as sometimes a menace to good government: sees voters as people who, in their own best interest, need to know as little as possible about a crisis – and the bigger the crisis, the less they should know.[35]

People made extensive use of the government's information hotline[36] the moment it was activated. The very fact that the hotline was soon overwhelmed with queries, revealed a much more serious problem: whereas contingency plans and clear assignment of responsibilities had been established in the event of a nuclear accident on British soil, there were no such precautions for a nuclear accident overseas.[37] Even the adequacy of existing response measures to incidents on British soil were now being questioned.[38] Press releases on 9 May that radiation levels in Scottish milk had indeed exceeded safety levels and that it was still unsafe to drink rainwater, did little to calm the growing anxiety.[39] The most obvious evidence of this confusion and anxiety in large portions of the British population were the approximately 800,000 phone calls to various agencies seeking information.[40]

In France,[41] late in the evening of 28 April, France 3, in its news show *Soir 3*, was the first TV channel to transmit the *TASS* press release to a French audience.[42] To gain more information, the journalists contacted their correspondent in Moscow, who was not able to add much to the original *TASS* press release, only some data regarding the nuclear plant. The staff of the Swedish power plant Forsmark, since they were the first in Western Europe to discover the radioactive fallout from Chernobyl, provided more information. French newspapers published the first articles about the accident on the morning of 29 April. The national daily social-democratic newspaper *Libération* informed its readers that, according to news agency *TASS*, an accident at the Chernobyl plant had caused victims and abnormally high levels of radioactivity had been measured in Scandinavia. With regard to the situation in France, the article went on to state that there was no cause for alarm, as the measurements taken thus far had not shown any abnormalities.[43] A similar all-clear appeared in the conservative *Le Figaro* and the far-left *L'Humanité* newspapers, revealing that initial reporting on Chernobyl did not depend on the political orientation of the newspapers. *Le Figaro* informed its audience on 30 April that another radioactive mass of air was approaching Poland and Germany but according to the Service centrale de protection contre les rayonnements ionisants (SCPRI), no radioactivity had yet been measured in France.[44] The same day, *L'Humanité* declared, quoting the CEA, every care had been taken to ensure no similar accident could happen in a French nuclear reactor.[45] Thus, the alienation of the problem, through reassurances to the French public that the 'nuclear cloud' had not yet touched their country and that the French plants were safe, was an intrinsic element of the first French news reports. Interestingly, the articles in *Le Quotidien de Paris* shifted their perspective from one day to the next: on 29 April the newspaper mentioned a slight design similarity in some USSR reactors with the West's PWRs.[46] The following day, however, it backpedalled, emphasizing that the Soviet nuclear fleet had nothing in common with Western plants. It then added that the USSR, in general, was incapable of preventing and controlling such accidents.[47] French journalists often linked such statements with national nuclear policies. For example, *Le Quotidien de Paris* stated that it was ridiculous to call into question the nuclear programme as such, just because of this 'set-back in progress'.[48] No newspapers were published on 1 May because it was a national holiday. But on 2 May, the reporting continued as it had begun, and the French public was assured once more that Chernobyl posed no threat to the country. *Le Matin* declared 'there is zero threat': thanks to the current meteorological conditions, France was protected from the

Chernobyl fallout: 'The national meteorological service forecasts that the anti-cyclone, this high pressure area which reigns over France right now, protects us from any arrival of a radioactive cloud'.[49] The newspapers also reported on the 'overreactions' of other countries, as did *Libération* in its article on children in Poland being intoxicated by iodine due to their parents' panicked reaction.[50] French parents were warned not to follow this example: according to Pierre Pellerin of SCPRI, there was no need for any countermeasures.[51] In general, journalists contrasted the chaotic reactions in other countries, especially West Germany, with how calmly and professionally the situation was handled in France. The journalists were thus relying on the evaluations and statements by the official experts they consulted on this nuclear emergency: SCPRI, CEA and the national meteorological service.

Public Authorities

Representation in Early Media Reporting

In 1986, the main institutions providing official expertise on radiation[52] protection and nuclear safety in France were the SCPRI and Institut de protection et de sûreté nucléaire (IPSN).[53] In late April 1986, the government and French journalists turned to these institutions for an informed evaluation of the risks the Chernobyl fallout posed. Their statements thus featured prominently in the newspapers and their representatives were invited to TV talk shows and for interviews. The SCPRI was responsible for measuring, observing and evaluating radioactivity levels and their impact on the French public. The SCPRI, which reported to the Ministry of Health, was founded in 1956 to create an agency responsible for radiation protection outwith the CEA, the French national agency for nuclear research and development for civil and military applications.[54] As the SCPRI expanded over the years, it was given more and more responsibilities. By 1986, it had become the national authority in charge of radiation protection, maintaining a network of monitoring stations to track radiation levels in the air.

The French public became familiar with Pierre Pellerin, SCPRI's founder and long-term director, through his public statements on the impact of Chernobyl in the spring of 1986. From then on, his name was closely linked to the *affaire Tchernobyl*.[55] Pellerin's involvement in this matter may have been one of the main reasons why, after his retirement, the SCPRI's activities were terminated (in 1994) and its responsibilities taken over by the newly instigated Office de protection contre les rayonnements ionisants (OPRI).

The other main French institution for radiation protection and nuclear safety was IPSN. Founded as sub-division of the CEA, and in charge of *protection et sûreté nucléaire* (nuclear protection and safety), IPSN also conducted research and provided expertise. The close ties between the CEA, the principal institution in the French nuclear sector, and the IPSN were highly criticized for constituting a lack of independent control in the nuclear sector. Consequently, a reform in 2002 restructured the French radiation protection and nuclear safety institutions. OPRI and IPSN were merged into one federal agency operating independently from the CEA: Institut de radioprotection et de sûreté nucléaire (IRSN).[56]

Back in 1986, however, Pierre Pellerin and his SCPRI were the main source of information regarding the intensity of radioactive fallout from Chernobyl in France. Parts of SCPRI's series of communiqués were reprinted in the newspapers. Members of the French government, such as the Minister of Environment, Alain Carignon, also based their public statements on this state agency's evaluations. The SCPRI did not communicate exclusively through its communiqués. It also sent its representatives to give interviews to newspapers and on TV. This direct communication is of particular interest here because the way in which the official nuclear experts had acted in public played a decisive role in triggering and shaping the *affaire Tchernobyl*. The story of the communiqués is history in itself given that, over the course of the *affaire Tchernobyl*, many French journalists claimed that Pellerin and the SCPRI had intentionally withheld the true figures regarding the fallout and not communicated all their data and insights to the media, an accusation they both rigorously refuted.[57]

The first time the SCPRI began with public interventions was when Pierre Pellerin came on TV for an interview with channel TF1 at noon on 29 April. When asked for his evaluation of the situation in Sweden regarding public health risks, Pellerin responded:

> I had information this morning directly from the Scandinavian [radiation protection] services that work with us within the framework of the international centre I am directing. These are very thorough people, and they tell us exactly this: yesterday, the iodine-131 reached 10 Becquerel per cubic meter of air and this morning it went down to below 2.5 Becquerel per cubic meter of air. This is a radioactivity that is noticeable, that is measurable, but it does not represent any inconvenience with regard to public health. There has been so much doom mongering with regard to nuclear that one risks unleashing panic. I would like to say it here, clearly, that even for the Scandinavians, there is absolutely no menace to health. It is a phenomenon that we are all observing, all the competent people in this field in Europe and in the world, a phenomenon which is very interesting to observe because of the lessons we will learn regarding the movements of air masses, masses of air ra..., of eventually radioactive air resulting from an accident, but that does not threaten anybody at the moment, except perhaps in the immediate neighbourhood

of the plant and, furthermore, it is foremost in the plant that the Russians admitted there were people who had been injured.[58]

Pellerin's interview was characterized by his cold, arrogant attitude towards his interviewer, who presented Pellerin as an ivory tower scientist, interrupting his speech to demand: 'En clair, en clair, en clair!' ('In plain language!'), an interruption to which Pellerin paid no attention whatsoever.

The SCPRI did not just comment on aspects of radiological protection. It also provided evaluations on nuclear safety issues. For example, in an interview published in *Le Parisien* on 30 April, a SCPRI representative stated: 'a major accident like the one in Chernobyl just cannot take place in France because of the difference in design that exists between the plant concerned and the type of plants which we build. … Our quality, safety and maintenance controls are a lot more rigorous than those in the USSR'.[59]

Just like the SCPRI, the IPSN made statements on Chernobyl in the early days of the accident, its representatives had a media presence and it provided evaluations of both nuclear safety and radiological protection. On 29 April, the IPSN director François Cogné gave an interview on channel Antenne 2. When asked if something similar could happen in France, Cogné replied: 'With regard to secrecy, I don't believe so. The electric power plants are totally controlled by the public and open to all controls'. And in response to the interviewer asking whether 'there is a risk these winds will reach France', he stated: 'Measurements have been taken … these measurements are completely negative and there isn't any reason, just in terms of the meteorological conditions, that whatever it may be would be measured in France'.[60] Cogné even provided a statement on the Chernobyl death toll – at a time when there was almost no information available on the details and evolution of the accident. Nevertheless, Cogné declared: 'the only deaths are probabilistic deaths. … At this moment, the only consequences are probably consequences for the workers in the plant itself and not for the neighbouring inhabitants'.[61]

An analysis of these early public statements on Chernobyl by French official experts reveals that the categories of comparison central to this study – personal affectedness, radiophobia and apocalypse, anti-East European/anti-Soviet stereotypes – all featured prominently in their narratives. With regard to personal affectedness, the SCPRI and IPSN representatives emphasized that there would be no fallout of any proportion that could pose a threat to French public health. Furthermore, they stressed that the Chernobyl plant had nothing in common with French nuclear power plants and what is more, their safety and security standards were superior to the practices applied in the USSR. These were the two points on which they based their reassurances that there was no reason to worry a similar accident could occur in a French plant. Within the framework of personal affectedness, this strategy of alienation aimed to prevent the French public from transferring the Chernobyl scenario to the French nuclear fleet. This element was underpinned and reinforced by references to anti-Soviet stereotypes, which relied heavily on images of Western (technical) superiority and Eastern unsoundness. The public experts' evaluations included statements on the situation in the vicinity of the plant. They strongly dismissed speculations of a death toll of 2,000 people circulating in the US media. French state officials were convinced that panic-induced overreactions fuelled by nuclear fears posed a greater risk than the fallout itself. Their narratives belong in the category of radiophobia narratives that aimed to oppose the narratives describing the situation as an apocalypse.

In Britain, statements by official radiation protection and nuclear safety experts also formed an essential part of the early media reporting on Chernobyl. The National Radiological Protection Board (NRPB) – basically the British counterpart to the SCPRI in 1986 – was the main source of information about radioactivity levels in Britain.[62] And, like their French colleagues, the British public experts initially saw no reason to worry about potential radioactive fallout from Chernobyl in their country. On 29 April, the *Guardian* wrote:

> Britain has no need to fear the radiation, the Protection Board said. The fall-out reaching Scandinavia was only twice the natural background level of radiation. This was so low that it should not cause any harm if it was ever carried to Britain. There was no need for any special precaution to be taken in this country. … Close to the plant the radiation doses may be lethally high but the amount of radioactivity so far detected in the west presents no immediate hazard especially as the isotopes are apparently of a kind which does not accumulate in bone or other tissues.[63]

A couple of days later, this narrative had still not changed. An article in the *Guardian* on 3 May quoted a representative of the official radiation protection institute: 'Since the NRPB has already checked nearly 100 people from places known to be affected by the accident – that is Minsk and Kiev – it is clear that people in Britain will not be placed at risk by the cloud'.[64]

The British government also gave an all-clear very early on. In an article on 30 April, the *Guardian* reported on the House of Commons debate held the previous day, when 'MPs were assured by both the Prime Minister and the Environment Secretary, Mr Kenneth Baker' that 'Britain has escaped the effects of the nuclear plant disaster in the Soviet Union'.[65] However, the article included critical voices in this debate, namely Tony Benn, Labour MP for Chesterfield, who called for a nuclear power phase-out and a full debate on nuclear power in Britain. Tony Benn specifically placed the Chernobyl event in relation to the nuclear policies regarding the ongoing new build project: 'He demanded that the Government give a full report on the Chernobyl incident before any authority was given to proceed with a pressurized water reactor at Sizewell'.[66] But the Thatcher government saw no reason why the Chernobyl accident should influence British nuclear policies. As the Prime Minister herself declared: 'The reactor in the USSR is totally different from that planned at Sizewell. The record of safety and design, operation maintenance and inspection in this country is second to none. I hope, therefore, you will think it right to support the furtherance of such an excellent nuclear industry'.[67]

Margaret Thatcher was certainly not alone in highlighting the differences between the Soviet reactor design and British plants: 'The energy secretary, Mr Peter Walker said during a visit to the Sellafield reprocessing plant that the Chernobyl disaster could not happen in Britain because there was no reactor comparable to the Soviet one anywhere in the western world'.[68] However, this narrative was strongly attacked by Greenpeace and Friends of the Earth. They said it was irrelevant whether or not the same thing could happen in the UK; the question was whether the consequences of an accident in the UK would be even more immediate, given that 'British reactors were not totally contained with a protective shell whereas the Chernobyl reactor had been'.[69]

Nevertheless, the government did not consider design issues to be the major obstacle to furthering its nuclear policies. The greatest obstacle lay in people's exaggerated perception of the risks associated with this technology. In the opinion of Energy Secretary Peter Walker, this perception of risk was not specific to nuclear power but inherent in 'anything that is tolerable, new and innovative. If I tried to introduce gas for the first time now, there'd be no possibility of having explosive gas going down every street and into every house'. In order to overcome this fear of nuclear power, he suggested following France's good example. An article in the *Guardian* quoted his reasoning:

> Instead of looking at the disaster in the Soviet Union where the design concerned had in any case raised widespread concern among experts, Britain should be taking its cue from France, where there was virtually no controversy about nuclear power. This, Mr Walker said, was because electricity was 20 per cent cheaper there – a direct consequence of the intensive use of atomic energy. He pointed out that 60 per cent of France's nuclear capacity was accounted for by Pressurised Water Reactors (PWRs) of the type which the CEGB wants to build at Sizewell. … [For Walker] there is no doubt that the only form of energy that is likely to give enough electricity at tolerable costs and safety is nuclear.[70]

Besides individual members of government, journalists turned to one person in particular to obtain information on Chernobyl's impact on Britain: John Dunster, the Director of NRPB. When Dunster died in 2006, the obituary published by the Health Protection Agency (HPA) – which NRPB had become part of – did not forget to mention the role Dunster had played in the direct aftermath of Chernobyl:

> John Dunster will also be remembered for the response of NRPB to the Chernobyl accident in 1986. The plume from Chernobyl arrived in the UK on Friday, 2 May, just before a bank holiday weekend. John Dunster led the NRPB response to the accident throughout that weekend and appeared on all the major television and radio networks to explain what the impact was likely to be in the UK. He understood the need for clear communications and drove forward efforts to get sensible advice on Ceefax and Teletext. There was understandable public concern at the time and John Dunster argued strongly that providing accurate public information about possible risks should be a priority.[71]

However, in 1986, some were critical of his public appearances. They were suspicious of Dunster because of his role in waste dumping at Windscale and in the evaluation of the Windscale Fire. An article in the *Guardian* on 7 May 1986 phrased this reservation as follows:

> During the past few days, Dr John Dunster, director of the National Radiological Protection Board, has done sterling work in calming public fears. Keep taking the milk but cut down on the rainwater, his latest reassuring message goes. Only 'a few tens' would die. This is the same Dr Dunster who reported in 1958 on a two-year ongoing experiment to increase radioactive waste discharges from Windscale deliberately in order to find out where the waste was going. … Four of the young leukaemia victims in the neighbouring village of Seascale, listed in the Black Report, were born in 1957 – also the year of the plant's fire. Dr Dunster subsequently replied that the discharges were properly authorised and within safety limits. In 1984 he urged a hunt for the 'missing factor' other than radiation that must be responsible for leukaemia cases in the area.[72]

Whatever this missing factor may be, or if indeed it exists at all, John Dunster's prediction regarding the impact Chernobyl would have on Britain proved to be wrong. In an article in the *Guardian* of 7 May he was quoted as saying: 'But if the cloud does not come back the whole thing will be over in a week or 10 days'.[73] This was definitely not the case, as the restrictions on sheep farms implemented over that summer would clearly show.

Comparing the early statements of French state officials with their British counterparts, it is first and foremost important to stress that representatives of the British institutions in charge of radiological protection and nuclear safety provided very similar statements to those pronounced by their French colleagues. Both dismissed any kind of personal affectedness, proclaiming that the radioactive fallout from Chernobyl would not lead to worrisome consequences in their country, nor did the event in the USSR give them cause to question their own nuclear programme. In addition, the experts who intervened in both countries did not appear in the media reporting as anonymous members of a public agency. They were personalized, namely Pierre Pellerin and François Cogné in France and John Dunster in Britain.

Early Reports

While the immediate reactions to the Chernobyl accident in late April and early May are crucial for understanding the tenor of the official response, considering only this time-frame would be reductive; it is far too brief to properly reflect on how the early narratives and interpretations developed and were communicated by the public authorities. Widening the time perspective enables us to consider the statements that actors initially transmitted in their newspaper and TV interviews and the statements the actors themselves published once they had obtained more detailed information, either by their own means or via the evaluations communicated at the international level, primarily through the IAEA.

At a conference organized by the British Nuclear Energy Society in October 1986 (the proceedings were published in book format in 1987),[74] NRPB representatives presented their evaluation of Chernobyl. Their statements offer interesting insights into NRPB's evaluation of the health impact. Regarding the 'Radiological Consequences in the USSR', M.D. Hill of the NRPB stated in his paper that there had been thirty-one deaths, adding: 'no one from the off-site area had to be taken to hospital for treatment of radiation injuries'.[75] Here, like many others, he drew upon the information provided by the report presented to the IAEA by the USSR delegates at the 1986 Chernobyl conference in Vienna. However, Hill criticized this very same Soviet evaluation regarding a different aspect of the accident, namely the consequences of the collective dose received by the evacuated population. He was convinced that the Soviet estimate was an exaggeration and wrote: 'more realistic assumptions could produce an estimate which is a factor of 10 lower'.[76] Thus, his statement 'fatal cancers expected to occur in the 135,000 evacuees as a result of the external irradiation would be about 200',[77] provided a rather low estimate of the expected death toll. As for the wider population, the NRPB also estimated a low health impact of the Chernobyl fallout. Hill's colleague, J.R. Simmonds, wrote in his paper that, based on the NRPB's own assessment of the radiation dose received in Eastern European countries, 'in all cases the doses are lower than those received annually from natural background radiation'.[78] This evaluation of the situation in Eastern Europe is indeed far from an apocalyptic scenario.

Interestingly, the consequences of the fallout in Britain were not mentioned in any NRPB presentation. This topic was, however, raised in the final discussion, and the conference proceedings quote John Dunster as saying: 'I do not know what the results in lamb show at the moment, and I have no idea for how long the restriction is going to persist, but probably not much longer'.[79] The fact that the Director of the NRPB was not up to speed on the levels of radioactivity in sheep at this crucial time is quite remarkable, considering the intense impact of the Chernobyl fallout on British sheep farming. I will further elaborate on this impact below.

Only a few months after this conference, in January 1987, the NRPB submitted a *Preliminary Assessment of the Radiological Impact of the Chernobyl Reactor Accident on the Population of the European Community*[80] to the European Commission. The European Commission had commissioned the NRPB to conduct this report: 'The aim of the study was to review information on the environmental contamination measured in member states of the EC; to make a preliminary assessment of individual and population doses for each country; to make an estimate of the resulting health impact and to indicate the effects of the various countermeasures'.[81]

In the general description of the accident, the authors stated that 'a series of human errors, whereby safety systems were deliberately switched off and operating rules were ignored, brought the reactor into unstable condition'.[82] Regarding the health impact, the NRPB calculated that

> the expected number of additional thyroid cancers occurring within EC countries due to Chernobyl is estimated to be some two thousand, of which about 5% are expected to result in fatality. The number of additional fatalities from cancer of all types due to Chernobyl is expected to be in the region of [a] thousand. These extra cancers are predicted to occur spread out in time over a few decades following the accident. These estimates need to be seen against the background of cancers that would occur in the population even if the Chernobyl release had not happened. Over the next fifty years about thirty million people in the EC countries are expected to die from cancer of one type or another.[83]

Thus, NRPB concluded that 'it will be impossible to detect the health impact of the accident'.[84] Reframed using the analytical categories in this research, this statement asserts that no grounds for perceptions of personal affectedness existed in Western Europe.

The brochure circulated by the United Kingdom Atomic Energy Authority (UKAEA, commonly known as AEA) is another example of an early Chernobyl report published by British public authorities. In some ways, the AEA fulfilled a similar role in early British nuclear history to the CEA in France.

It was founded in 1954 as the principal authority of the British nuclear programme, for its military and civil divisions alike. It coordinated the research programmes and operated the sites. Soon after its foundation, some of the AEA's activities were outsourced and new institutions established. This led in 1971 to the formation of the NRPB and British Nuclear Fuels Limited (BNFL), a private sector company that was thereafter responsible for fuel production, including operating the Windscale site. Consequently, the AEA's field of activity shifted to decommissioning the old Magnox plants and to nuclear fusion research, its main area of operation today. Although no longer the central actor as at the time of its establishment, in 1986 the AEA continued to play an important role within the British nuclear sector and was considered the leading authority in nuclear research. In 1987, it published the brochure *The Chernobyl Accident and Its Consequences.*[85] Lord (Walter) Marshall of Goring, chairman of the Central Electricity Generating Board (CEGB) wrote the foreword.[86] In the 1980s, Marshall was a key figure in the British nuclear energy sector. He was appointed chairman of the AEA in 1981, knighted the following year and the year after moved over to the CEGB. His enthusiasm for nuclear energy generation was highly appreciated by the British government, namely Margaret Thatcher. For 'the British anti-nuclear movement [however] he occupied the top slot in their demonological lexicon for a decade and a half, as he advocated the virtues of pressurized water reactors and a major construction programme of nuclear power plants in Britain'.[87] His role in the aftermath of Chernobyl was so important, that even when he died ten years later, his obituary in the *Independent* stated that Marshall had 'bent all his powers of communication to explaining to professional and lay audiences what had gone wrong and what the essential differences were between the Russian and Western concepts of nuclear safety management'.[88] This commitment also came across clearly in his foreword to the AEA brochure, where he stressed that the British nuclear sector had to give 'as much information as possible to the British public in a language which is both correct and understandable'[89] in order to convince people of the national nuclear operators' trustworthiness.[90]

The AEA booklet covered a wide range of topics, including the USSR power reactor programme, the evolution of the accident, the Chernobyl source term and its consequences for the environment. The information was mainly taken from the USSR report to the IAEA conference in August 1986, and based on this evaluation, the brochure's authors reasoned that such an accident could simply never occur in the UK. The section dealing with the environmental consequences of Chernobyl was the shortest of all. The AEA delimited the area of Chernobyl's health impact exclusively to Eastern Europe and even presented a precise number: '7,500 fatal cancers in the European part of the USSR [have to be expected] as a result of the accident'.[91] This number was then statistically compared to other health risks. More specifically, the AEA chose one case in particular for the comparison, one that I will refer to as the 'smoking-topos'. The smoking-topos, the comparison between the yearly death toll from smoking and the Chernobyl death toll, was used to highlight how minor the health impact of the accident was. The smoking-topos, however, was not presented as a death toll in the AEA brochure but as a proportion. The 7,500 fatal cancers attributed to the Chernobyl accident were declared 'equivalent to the compulsory smoking of less than 3 / 10,000 of a cigarette per week for 30 years'.[92] Understanding the argument behind this comparison is particularly useful as it explains AEA's approach to the problem of Chernobyl-induced health effects. The authors of the brochure had summed up the health effects if each individual inhabitant of the European part of the USSR smoked the tiny amount of 3/10,000 of a cigarette every week for thirty years. The total sum of all these minute statistical health effects was 7,500 – the number of cancers statistically calculated to appear in Eastern Europe as a result of the Chernobyl fallout.[93] Right from the outset, public authorities used such comparisons of statistical data as major tools to 'put Chernobyl into perspective', illustrating that everyday life was rife with health threats far bigger than the Chernobyl fallout; and stressing that compared to other causes of death, the Chernobyl health impact was at best negligible. We have already encountered this narrative strategy in the NRPB report above, which related the calculated Chernobyl death toll in the EC to the general death toll from cancer.[94]

To extend the sources for building my argument on the AEA's early Chernobyl narrative, I found it opportune to include the 1988 AEA publication *Nuclear Safety after Three Mile Island and Chernobyl*, which evaluates the accident.[95] This book contains the proceedings of the June 1988 *International Approach to Nuclear Safety* conference held in Blackpool. The book's editor, G.M. Ballard, was Head of the National Centre of System Reliability, based at the AEA. The preface opened with a very reassuring statement: 'There are slightly different viewpoints around the world on some of the detailed technical areas, but the overall picture painted by the papers is one of a well considered and researched approach to ensuring that nuclear power plants are designed and operated to high standards of safety'.[96] However, the preface did not mention that the viewpoints on technical details were not just slightly different and that in fact profound disagreements had already sprung up by 1986 about the death toll. In the introduction, Ballard described the underlying frame of AEA's interpretation of Chernobyl's impact. In his opinion, there was absolutely no reason to question nuclear power as a result of the Chernobyl accident. This was because it 'offers the only viable, economically-competitive source able to provide secure supplies'[97] and there was simply nothing to worry about with this technology: 'The safety experts know nuclear power is safe but must bridge the gap between their views and the public perception'.[98] For Ballard, it became obvious how little the public perception of the risk surrounding nuclear energy was really grounded in truth, in the debate over low-level radiation, as 'all attempts to explain leukaemia clustering on the basis of radiation from nuclear installations have failed'.[99] He stressed that instead of posing a threat, nuclear power was 'environmentally benign', and that the 'nuclear community is intensifying its effort to get this message across to the public'.[100] It would become obvious in the 1990s that nuclear advocates had succeeded in positioning nuclear power as an environmentally benign source of energy, when the public image of nuclear power was largely transformed into one of a saviour from the threats of climate change, a process commonly known as the 'greening of the atom'.[101] But in 1988, shortly after the Chernobyl disaster, this positive image had yet to be (re)established. Ballard attempted this by pointing out that the 'emphasis on reactor safety has a key role to play in the future of nuclear power by helping to ensure that … the public are satisfied that the technology is safe and that they are completely protected'.[102] Given the importance the editors placed on this matter, the differences between the 'faulty RBMK design' and the 'safe design of British reactors' were greatly emphasized in the book. Thus, in his paper 'Analysis of the RBMK Against UK Safety Principle',[103] P.G. Bonell of the AEA presented a list of thirteen points showing that the RBMK and the UK safety principle were simply not comparable. Other papers concentrated on the problem of the man-machine interface or the protection measures against fires and earthquakes. One representative from France presented the EDF's safety policy as an example of good practice. Another AEA paper, 'Modelling the Consequences of Reactor Accidents',[104] underpinned the point made by Ballard, that the effects of low-level radiation on health were not a crucial topic – their list of fields in which research was being carried out in the US and the UK did not include the health effects of low-level radiation. When analysing the AEA narrative on Chernobyl, it is also important to consider the death toll presented in the paper 'Consequences of the Chernobyl Accident' by W. Nixon and M.J. Egan (both AEA employees). Although they had calculated 'a total number of cancer fatalities in Western Europe, arising from the Chernobyl accident over the next decades, of around 1500', this number was followed by the statement: 'this figure should be regarded as a likely upper estimate'.[105] The way the AEA (and most nuclear authorities in the years following the accident) calculated the Chernobyl death toll, was based on two main assumptions. Firstly, the models created using research on the health impact of Hiroshima and Nagasaki were applied to the Chernobyl fallout; which is to say, radionuclides released by atomic bombs and how they were taken up by humans, were equated with radionuclides released into the air by the burning reactor in Chernobyl. Secondly, a linear relationship between radiation dose and the risk of cancer mortality was hypothesized; this relationship was deemed 'conservative' and the resulting figures as probably be too high. In the years to come, these two assumptions became the focus of attention for actors criticizing the numbers presented by the official experts. Critics questioned these assumptions, claiming that completely new models were necessary to fully and accurately evaluate the health impact of Chernobyl. But before considering the critique levelled at these official Chernobyl reports, it is important to include the narratives provided by other British state officials.

The UK government, in an attempt to address and overcome the accusations from critics in early May 1986 that its response to Chernobyl severely lacked coordination and that it had not conveyed correctly all the information available on radioactivity levels in Britain, decided to adopt a proactive strategy. In July 1986, Her Majesty's Stationery Office (HMSO) published the report *Levels of Radioactivity in the UK from the Accident at Chernobyl USSR on 26 April 1986.*[106] It was jointly edited by all the public entities[107] involved in the post-accident monitoring and decision-making process and contained 'information available to NRPB up to and including 1986. The data are arranged by environmental medium and geographical location in chronological order'.[108] In practice, this meant pages and pages (almost 200 in total) of columns of numbers. The report explained in detail the methods used to procure these measurements such as how samples had been taken. However, this report did not include any analysis of the data. Consequently, rather than bringing true transparency to the decision-making process, this report was a tactic to show the public the amount of data on which public authorities had based their decisions, thus demonstrating that their decisions were grounded in science and as such valid.[109]

Publishing a series of numbers, however, was not enough to mitigate all doubts about the appropriateness of the government's response to Chernobyl, doubts that persisted not only in the House of Commons but beyond Westminster's walls. To gain further insight into the government's action in the aftermath of the accident, the House of Commons Agriculture Select Committee requested that the government write a report, which was submitted in the summer of 1988.[110] The Ministry of Agriculture, Fisheries and Food (MAFF) compiled a supplementary memorandum to the government's response.[111] Although this memorandum was not aimed at a particularly large audience, I refer to it nevertheless, as it is a very interesting source. It gives detailed insight into the tasks and problems facing the MAFF in the wake of the accident, particularly concerning the aforementioned sheep farm restrictions. In addition, the memorandum explained newly introduced EC regulations on radioactive food contamination and provided a detailed list of its post-Chernobyl research projects, for instance the 'Dynamics of Radionuclide Uptake in Sheep' and 'Effects of Minerals on Reducing Caesium Uptake and Distribution in Upland Pastures'.[112] Although the MAFF had encountered considerable difficulties with the sheep farm restrictions – ranging from identifying appropriate marking colours to compensating the farmers – its conclusion about the situation was fairly positive:

> The extensive monitoring and rigorous controls imposed within the UK have proved successful in retaining confidence in the safety and quality of UK foodstuffs at home and abroad and UK exports of foodstuffs have not been adversely affected. In particular British exports of sheepmeat rose by around 28% in 1986 compared to 1985. Maintenance of confidence in British produce remains a high priority for the Ministry.[113]

It is interesting to contrast this memorandum, not targeting a wide audience, with the British nuclear authorities' communications: their publications all omitted to mention the issue of contaminated sheep. For British nuclear safety experts, the link between the UK and Chernobyl constituted the necessity to improve the man-machine interface in a power plant, but not problematizing British Chernobyl fallout. For them, Chernobyl had demonstrated that the weakest link in nuclear technology was the 'human factor' in running the plants, not the risks the technology itself posed for humans and the environment in general. Along these lines, at a conference in London in December 1986, E.A. Ryder, Her Majesty's Chief Inspector of Nuclear Installations[114] stated: 'the main lessons from Chernobyl were the need to adopt a safe design and to rely more on layers of automatic control and protection equipment rather than on fallible human operators'.[115]

Much like their French colleagues, British state officials focussed their early Chernobyl narratives on three main aspects: the design differences between the RBMK and national plants, the low hazardousness of the nuclear fallout in Western Europe, and the limited impact to public health in Eastern Europe. Combining these aspects provided a narrative of Chernobyl that de-emphasized or dismissed the importance of the accident for the global nuclear programme. Undoubtedly, the accident could serve as an opportunity to conduct further research in radiation ecology and thus gain more information on the transport mechanisms of radioactive particles in air, water and soil. Regarding the health impact of these particles, however, French and British state officials were convinced that they had the means to calculate its impact (and thus the death toll), although they admitted there were some unknown variables in their equations. Although people would most certainly die because of Chernobyl, in relation to other threats to public health, such as smoking, these few deaths were not even observable. Yet, one outcome of Chernobyl was very clear to French and British nuclear state experts, who had identified a new task for their work agenda: they needed to better educate the panicking public about the safety of nuclear installations and the benignity of the nuclear enterprise.

Nuclear Power Industry

Having analysed the French and British authorities' early interpretations of Chernobyl, I now turn to the early narratives by actors connected to the nuclear power industry. The dividing line between these two actor clusters, especially with regard to 1986, is actually quite artificial, as all major institutions and companies in the nuclear sector were state-owned at this time. But because this analysis traces these actors and their narratives over the ensuing twenty years, it makes sense to apply this differentiation from the start. The nuclear sector in France as well as in Britain underwent profound reorganization and privatization in the years following the Chernobyl accident. For this reason, I discuss Public Authorities and Nuclear Power Industry clusters separately. Due to its strong dependence on state funding and guarantees, the nuclear sector has always been closely linked to the state. The categories I introduce here are therefore a means of structuring my material as coherently as possible rather than statements about public-private interdependencies in the nuclear enterprise.[116]

In the case of France, statements made in 1986 by plant operator EDF could very well come under the category of Public Authorities. The establishment of the EDF in 1946 as the national public electricity producer and provider, meant that the multitude of local producers and providers were socialized and centralized under one overarching operator. The EDF's nuclear history was marked for a long time by its rivalry with the CEA within the context of the *guerre des filières*, the struggle over which reactor design should be implemented on a large scale in France.[117] Due to its size – in 2008, the EDF had 104,000 employees – and its important decisively political role, the EDF has been often characterized by its critics as *toute-puissance* (all-powerful) and *Etat dans l'Etat* (a state within the state). In 2004, the company became a stock corporation. Almost all the stocks, however, have remained in public hands. In 2008, approximately 85 per cent of the stocks were still direct assets of the French state.[118]

Back in 1986, the EDF was not just a state-owned company, but its representatives were also attributed the role of public authority. Pierre Tanguy, the EDF's inspector general for security and safety, gave interviews and figured prominently in the media like his colleagues Pierre Pellerin and François Cogné. In an interview on 30 April 1986 during a channel Antenne 2 news show, Pierre Tanguy stated: 'If it [the "nuclear cloud"] will arrive here, I think one should ask the weather specialists, but the toxicity … that absolutely does not represent any danger … that becomes totally insignificant'.[119] Interventions by EDF representatives in the early media reporting on Chernobyl were not the only statements released to a broader public. In October 1986, the company published a *dossier d'information* recapitulating the topics discussed at the IAEA conference in Vienna.[120] Like the British AEA brochure, it claimed 'it is possible to take a more or less complete picture of the causes, the evolution and the consequences of the accident'.[121] There were still questions the Soviet attendees had left unanswered, but this information would soon be provided. However, the causes of the accident were clear, they 'can be found in its design "deficiencies", but at the same time, it was provoked by an incredible sequence of human error'.[122] Like the statements released by French public authorities, the EDF's dossier emphasized the differences between France and the USSR with regard to reactor designs, approaches to safety and operating conditions. Furthermore, the EDF drew parallel conclusions to the AEA regarding lessons to be learned from the Chernobyl incident, namely that the possibility of human error in plant operations had to be reduced. In addition, the public needed to receive information on nuclear issues in a more comprehensible manner. So, while the British nuclear experts emphasized educating the public, their French colleagues, rather than passing on more information, felt that the already accessible information had to be made easier to understand. At the Vienna conference, in an interview with *Le Quotidien de Paris*, Jacques Leclercq, the EDF's *responsable de la production* (Head of Production) announced that 'from now on, we will work together with communication specialists'.[123]

One of the first items in this promised communication campaign was the *Tchernobyl* brochure, published in 1987.[124] The aim to make the information easy to understand was clearly addressed on the first page, where the reader was told that the EDF intended to 'simply re-explain, in a way that would be comprehensible for everybody, these regrettable events which should not have a successor after Chernobyl'.[125] Like the early publications disseminated by British public authorities, the authors of the EDF brochure considered a comparison between the risk imposed by Chernobyl and health risks encountered in daily life as the most suitable way to put the accident's impact into perspective. Thus, the first page compared people's nuclear fears today with the fears of former generations towards the first railways and aeroplanes; the brochure added that this comparison was of course misleading, given that trains and planes had 'caused infinitely more deaths than Chernobyl'.[126] The first section of the brochure gave a general account of nuclear energy. To ensure everyone could understand the narrative, it provided the basic vocabulary, along with descriptions of the various kinds of ionizing radiation and basic physics of how nuclear plants functioned. The text then addressed Chernobyl, describing the accident as a result of 'human error, an unqualified operation crew … and a less stable reactor than ours'.[127] This narrative strategy of alienation was underpinned by references to anti-East European/anti-Soviet stereotypes, implying that plant workers were less skilled and reactors less well designed in the East than in the West. In terms of evaluating Chernobyl's health impact in Eastern Europe, the narrative was far from apocalyptic. The brochure stated that apart from the firemen who were fatally irradiated during the emergency response, there would not be 'any short-term deaths'.[128] Moreover, the long-term health impact would be minor: '300 supplementary fatalities over the next twenty years within the totality of the evacuated population (135,000 inhabitants)'.[129] The topic of health effects in France was only mentioned once in the forty-page brochure: 'One can say that in France, the general radioactivity was a hundred times lower than during the atomic [weapons] tests of the [19]60s, and that the most exposed citizen will not experience more damaging health effects than were he to have undergone one or two pulmonary x-rays'.[130] Hence, the EDF rejected entirely the issue of French personal affectedness regarding the transfer of the accident scenario to French plants as well as the possible health effects of the fallout. Instead, the emphasis was on radiophobia: the comments on general health effects ended with: 'the political and journalistic exploitation of this phenomenon [of very low doses of radiation] contributed to creating panic in certain countries, in particular in Germany'.[131]

Another strategy the operators in both France and Britain adopted to rebuild public trust in the national nuclear fleet, was to encourage direct interaction with the technology itself. Before the accident at Chernobyl, many sites already had information centres, and guided on-site visits had been a popular weekend family activity for quite some time.[132] This kind of direct interaction with the interested public intensified after Chernobyl.

Similar to its French counterpart EDF, the CEGB played an active part in providing the broader public with a firsthand expert evaluation of Chernobyl's impact. Until its privatization in the 1990s, the CEGB was the key player in the UK's electricity industry, responsible for electricity generation in England and Wales; it owned the nuclear and other power plants in these parts of Britain, and the grid itself. Furthermore, it served as a research institution. The CEGB and the EDF held similar positions in their respective national nuclear sectors. However, in the following decades, the privatization of the EDF would consolidate its role in France and even transform it into a global player, whereas the CEGB would be broken up by the privatization of the British electricity sector – a process, incidentally, during which most of the British nuclear power plants were taken over by the EDF.

In September 1986, a month before the EDF's brochure, the CEGB published its information dossier *Chernobyl*.[133] This spoke of the RBMK reactor design, how the accident occurred (claiming the faulty design and the operators responsible) and the containment measures, but also the fallout that had spread over Europe and its health impact. With regard to the current death toll, it stated: 'some 31 people have died as a result of the accident either directly or as a result of receiving lethal radiation doses'.[134] The long-term health effects were also discussed and directly compared to 'natural background cancers' – alongside the smoking-topos, this was another common topos to put Chernobyl's health effects into perspective. The authors, J. Collier and M. Davies, elaborated this comparison: 'Current estimates indicate perhaps 6–40,000 thyroid cancers and other cancers resulting over the next 40 years in the affected parts of Russia and Western Europe. This figure needs to be compared to a figure of 40 million cancer deaths expected in the same population over the same time period'.[135] Thus, they declared that the immediate and longer-term health impacts of the accident would be limited. On this aspect, the CEGB narrative was similar to what the EDF would provide. The aspect of anti-Soviet stereotypes, however, was a different matter. The CEGB actually praised the Soviets' management of the emergency. Describing the situation in the vicinity of the reactor as far from chaotic, the authors concluded the brochure with the statement: 'Given that no accident of such a magnitude had previously happened to any nuclear power plant in the world, the co-ordination and response of the many Soviet recovery services appears to have been exemplary'.[136] Many accounts had placed these emergency operations under a different light, speaking of a general irresponsibility of the Soviet leaders towards the population. This criticism was underlined by the claim that Pripyat had been evacuated far too late and that the firefighters had been sent in to get the fire in the burning plant under control without having first been properly warned about the life-threatening risk of radiation. Although this brochure did not use anti-Soviet stereotypes to frame the emergency operations, the CEGB did use them on other occasions to explain the accident. In a paper presented by B. Edmondson, Director of the CEGB Nuclear Co-ordination Group, at the conference *Nuclear Risks – Reassessing the Principles and Practice after Chernobyl* held in London in December 1986,[137] he attributed the accident to a kind of simplemindedness of Soviet nuclear engineers and plant operators. In his paper 'The Chernobyl Accident and Its Implications for Operators of Civil Nuclear Power Plants in the UK', Edmondson stated that 'the RBMK designers had suffered a tremendous psychological blockage in not foreseeing that the plant operators would commit an extensive series of violations of vital safety procedures'.[138] Regarding the safety of British nuclear power plants, on the other hand, Edmondson saw no cause for concern given that 'the RBMK reactor design is very different indeed from anything elsewhere in the world'. For him, this fact 'by itself could perhaps provide a satisfactory and resounding "No"' to the question 'Can it happen here?'[139] Thus, Edmondson's narrative contained a strategy of alienation, putting the responsibility on an entity alien to the British nuclear programme. However, seeing as 'operator action is a common threat with all reactor types',[140] the UK was, from Edmondson's point of view, not completely free of risk from possible future accidents. Therefore, he proposed taking action in order to guarantee operator training 'in the deeper "educational sense."'[141] This CEGB nuclear expert did not consider the technology in itself dangerous. The employees posed the true problem. Like the AEA, the CEGB called for improvements in the man-machine interface.

This narrative of placing the blame for the accident on a technical setting alien to the British nuclear sector, and on unqualified East European operators, was also espoused by CEGB chairman, Lord Marshall of Goring. In his contribution to the House of Lords debate *Nuclear Power in Europe* on 20 November 1986 – later published by the CEGB as a leaflet for wider distribution – he explained why 'Chernobyl could not happen here'.[142] Marshall reached this conclusion using the 'Russian' delegation's report submitted to the IAEA conference in August. The shortcomings of the reactor design had caused the accident, shortcomings that did not exist in British plant design and safety culture, and that had, as it happened, already been 'identified by a team of British engineers who studied this reactor in 1975'.[143] One crucial shortcoming in the RBKM design, was its *positive void coefficient*. This system specificity required operators to take an active role in the decision-making process in the event of an unstable situation in the reactor. Western nuclear engineers drew a line between 'Western' safety culture based on built-in protection, versus the 'Soviet' safety culture, which, from their point of view, gave far too much responsibility to the plant operators. Based on these assumptions, Marshall reasoned: 'We are satisfied that there is no narrow technical issue which we in the West could or should learn from the disaster. We have very well established safety rules which we follow meticulously. The Russians have chosen to ignore many of them and sadly have now paid the price'.[144] Thus, there was no reason to let the Chernobyl accident influence the decision on reactor new build projects in Britain, namely the Sizewell B project. After all, Britain had to keep up with other nations in this important economic field: 'Our competitors in France and Japan have reaffirmed their intention to expand their nuclear programme, and the Russians themselves have recently outlined ambitious plans to increase nuclear power production'.[145]

The CEGB's reasoning was replicated by another key player in the British energy sector, the Watt Committee on Energy. This committee, instigated in the aftermath of the 1973–74 oil crisis, combined 61 British engineering and research institutions, ranging from the British Nuclear Energy Society to the Royal Institution of Naval Architects.[146] In 1988, the Watt Committee on Energy published the report *The Chernobyl Accident and Its Implications for the United Kingdom*.[147] The report was compiled by an internal working group[148] and consisted of eight sections, each written by different committee members, with an additional ninth section on recommendations and conclusions by the entire working group. Funding for the publication of the report came from the CEGB, the South of Scotland Electricity Board, British Nuclear Fuels, the UKAEA and the National Nuclear Corporation. The aim was to reach and inform a broad audience on Chernobyl, not just people linked to the nuclear sector. According to the Watt Committee, its role was 'to promote and assist research and development and other scientific or technological work concerning all aspects of energy; to disseminate knowledge … [and] to promote the formation of informed opinion on matters concerned with energy'. Therefore, by publishing this report, it clearly intended to claim the role of an independent actor in the Chernobyl debate, stressing that it represented 'some 500,000 professionally qualified people', the UK's entire community of energy experts.

In his foreword, Watt Committee chairman G.K.C. Pardoe hinted that for the committee, the accident had a more political than health impact: 'There are comparatively small numbers of people, mainly in limited areas, for whom the direct effects of Chernobyl are still important, and the long-run indirect effects are important for almost everybody'.[149] Particularly interesting parts of the report's narrative on Chernobyl are: section 4, 'The Radioactive Release from Chernobyl and Its Effects' by B. Smith and A. Charlesby; section 5, 'Accident Management in the USSR and the United Kingdom' by G. Lewis; as well as section 9, 'Comments, Recommendations and Conclusions'.[150] The authors of section 4, after discussing the inventory of the release, its deposition to the ground in the UK, and the contamination of foodstuffs caused by the deposits, commented on Chernobyl's health effects. Their remarks are worth quoting:

> Beyond 100 km from Chernobyl, the effects of the additional uptake of radioactivity into the human population are likely to be so small that they will be impossible to detect even by the most careful of medical surveys over the next few decades. This simple truth was largely ignored by the media and has led to considerable anxiety and exaggerated fear of the nuclear industry. On the other hand, some deaths may well result, although how these should be interpreted is a matter of some debate: for example, many may occur in old people nearing the end of their lives anyway. This is not intended to sound callous; but it is intended to put the problem in perspective. W.K. Sinclair has given statistics of deaths per year in the US from avoidable 'accidental' causes. Nuclear power generation, including the very occasional release, is said to cause typically 100 deaths per year. Smoking causes 150,000, alcohol 100,000, road accidents 50,000 and accidents with guns 17,500. Chernobyl almost fades into insignificance by comparison.[151]

Leaving aside the problem of a comparison in which, as the authors themselves stated, the exact figures on one side are not known, it is difficult not to find this statement 'callous' or biased. For the most recent number of Chernobyl fatalities, the Watt Committee report listed two plant workers who had died in the explosions and resulting fires, and twenty-nine others who died in the following days due to radiation exposure. Regarding the health impact on the 135,000 evacuees, the report stated that 'none of these people showed any clinical symptoms, although it is estimated that up to 1,000 of them may develop cancers in the next few decades'.[152] For the health impact in the UK – which, according to their calculations, would amount to eighty thyroid cancers, four fatal cancers and forty general fatal cancers over the next fifty years[153] – the report presented a comparison that can be considered yet another topos in the Chernobyl debate: comparing the Chernobyl fallout to natural background radiation. The authors wrote: 'A much greater risk affects people living in areas with high radon concentrations, for example in Cornwall. Well-sealed modern housing in such areas can, it is believed, give rise to a risk of 1 in 2,500 of developing associated lung cancer in each year. This is some 10^4 times greater than the risk from Chernobyl fallout'.[154] The topos of the 'Chernobyl fallout vs. natural background radiation' comparison can also be found in the French discourse, where in terms of a problematic area with high radon background radiation, the Massif Central can be equated with Cornwall in South West England. The intrinsic flaw in this comparison is the problem of an unweighted comparison: whereas the exposure to radon in Cornwall is a natural environmental condition, the Chernobyl fallout was man-made and could have been avoided. If the Watt Committee's statements on the health impact are analysed in terms of the analytic categories of personal affectedness and radiophobia/apocalypse, the report clearly argued against exaggerated nuclear fears, in Western as well as in Eastern Europe. Though Chernobyl had produced an undeniable health impact, as the authors themselves wrote, it 'faded into insignificance by comparison' to other risks.

An even more interesting aspect of the Watt Committee report is its description of how the Chernobyl accident was handled in the USSR (section 5). The entire evacuation process was presented as having been well organized and executed in such a way to ensure the accident would not cause any harm to the people in the region. For example, the schedule of evacuations – severely criticized in many other accounts, especially by authors within the USSR – was described as rather well planned: 'The delayed evacuation of the population in the vicinity of the NPS [Nuclear Power Station] is understood to have been in accordance with official Soviet evacuation plans, which closely followed the recommendations of the International Commission of Radiological Protection'.[155] According to the Watt Committee report, the evacuees were well taken care of, as 'the relocation centres were equipped with medical and other emergency service resources to carry out personal decontamination, compulsory dosimetric monitoring, blood sampling for laboratory testing and replacement of contaminated clothing'.[156] Likewise, the procedures to extinguish the fires and build the sarcophagus were described as good examples of organization skills: 'Dose limit controls meant a large replacement work force over a considerable period of time – a formidable task'.[157] The word 'liquidator' was not mentioned once in this account. The only time the work force was personalized, was when they were called 'volunteer miners'.[158] However, when assessing this Chernobyl narrative using my third category, this account cannot be described as pro-Soviet. This positive account of how the accident was contained and managed by the Soviet authorities was intended to illustrate the manageability of a nuclear accident. My hypothesis is that the reasons underlying this narrative should be interpreted as: if the Russians can handle such an accident, there is even less reason to worry that an accident might occur in a Western plant.

This narrative of a manageable accident reinforced the narrative of easily calculable health risks. Altogether, Chernobyl was presented in the Watt Committee report as a conventional industrial accident. Accordingly, the 'Comments, Recommendations and Conclusions' (section 9) pointed in a very clear direction. For instance, it proposed that: 'nuclear experts from the industry should be encouraged to provide specialist lectures to graduates and undergraduates as well as, perhaps, to schools'.[159] Here again, it was the public's ignorance of scientific facts that had caused a problem, not the technology as such. Another disturbing factor had to be tackled: the plant worker. The question whether 'such an accident could occur in the United Kingdom' was answered with a 'no' – as long as there was a guarantee that 'failure of the human system' would be avoided.[160] 'Reactor Operation and Operator Training in the United Kingdom' (section 7 of the report) were thus seen as the major fields requiring action. After all, 'the accident would not have occurred if the plant operators had proceeded to carry through correctly the special instructions necessary for the operation of the test'.[161] A second strategy of alienation becomes obvious in this account: the scientific experts in the British nuclear sector not only drew a clear line between their own expertise and knowledge in reactor design and nuclear safety, and that of their colleagues in the USSR. In addition, they clearly distinguished between scientists and plant workers. In both instances, the fact that such distinctions were made at all denoted the inherent belief that the others, in other words the Soviets and the plant workers, were inferior.

However, this distinction between the 'knowledgeable scientist' and the 'untrustworthy operator' was not exclusive to British narratives. In France as well, the man-machine interface was identified as the area in which French nuclear experts saw room for improvement. Aside from these two national cases, the 'failure of the plant operators' along with the 'faulty design', became the universal narrative for explaining the cause of the accident. During the 1986 IAEA Chernobyl conference in Vienna, the USSR delegates had declared – Western accounts on this conference often used the term 'admitted' – that the accident had been caused by the irresponsible actions of the plant personnel. It was this argument that led to the 1987 trial against six Chernobyl plant technicians and officials – often referred to as the 'scape-goat trial'. It did not help the accused that though 'three defendants blamed the reactor's design or equipment for the disaster, Soviet officials said the accident, which occurred April 26, 1986, was entirely a result of human error'.[162] Over the coming years, this interpretation would guide risk management research in the nuclear industry to focus on how to better train staff to trust the machines and not their own evaluations in an emergency.

One more institution in the British nuclear power industry that took an active role in the early Chernobyl debate, deserves a mention: The British Nuclear Energy Society (BNES). The BNES was a learned society for professionals working in the nuclear sector. In 2009, it merged with the Institution of Nuclear Engineers to form the Nuclear Institute. Like many other publications on Chernobyl by British nuclear actors, the appraisal BNES published of Chernobyl in 1987 consisted of a compilation of papers presented at a conference, in this case: the BNES conference in London in October 1986.[163] Representatives of the nuclear sector and public authorities – namely the NRPB and the AEA – took part. The opening address by BNES Director General, J. Rimington, expressed his stance on the impact of the accident: 'The principle immediate effect of the Chernobyl accident is political and throws the future of nuclear power generation into the political cauldron'.[164] It is interesting to note that the BNES only started to consider nuclear power generation a political topic after the Chernobyl accident. This statement indirectly reveals how minor a role the anti-nuclear power plant protests had played in the nuclear history of Britain.

The British nuclear energy actors' early evaluations of Chernobyl's health impact reveal a salient point: that they deemed the health effects to be quite limited. They also clearly accepted as true the IAEA 1986 conference's USSR narrative regarding the immediate victim count and evolution of the incident and presented these details as uncontested facts. In those instances where they were critical, it was because they expected the health effects around the plant and in Eastern Europe to be much lower than estimated by the USSR delegates. On the cause of the accident, these narratives willingly accepted the explanation that it had been a combination of factors, the faulty reactor design and the violation of clear safety rules by plant personnel. From this standpoint, the British actors were able to build their argument that such an accident could simply not happen in the UK. If the British nuclear sector was to learn any lesson from the Chernobyl accident, it was that the human element had to be contained as much as possible and the mechanical procedures given precedence. Furthermore, the narratives pushed for increasing communication with the concerned public, or preferably improving the education of this misinformed public the moment the reactions to Chernobyl showed how many people were needlessly worried about ionizing radiation. This situation had to change in order to ensure the good side of nuclear energy would be appreciated, not least because of future expansion in this technology. Compared to the advantages of nuclear electricity generation, the impact of the Chernobyl incident, in the short as well as in the long term, was relegated to the position of just another industrial accident.

Critical Voices

Following heavy rainfall in early May, the MAFF started collecting foodstuff samples to monitor radioactive contamination. Aside from radioactive iodine, which has a short half-life and therefore ceased to cause concern already after several days, another radionuclide with a longer half-life needed to be monitored: caesium. Levels of radiocaesium remained high in one type of foodstuff in particular: upland sheep. Many of the meat samples collected by the MAFF exceeded the newly established EC-threshold of 1,000 Becquerel per kilogram. Therefore, restrictions on the movement, sale and slaughter of sheep were imposed: in England and Wales on 20 June, in Scotland on 24 June, and in Northern Ireland on 14 September.[165] Over the following months, several compensation schemes were introduced, and a mark and release system was implemented for the sheep in restricted areas. Initially, the restrictions were imposed on nearly 7,000 farms, so 4 million sheep or one-fifth of the UK sheep population.[166] The farmers were told that these restrictions would only be for a couple of weeks: the radiocaesium would soon be washed into deeper layers of the ground or locked into the soil and thus not be taken up by the pasture, and consequently not by grazing sheep. However, this was not the case: levels of radioactivity remained high and, although the ban was lifted for most of the restricted farms over the next three months, many farmers would have to live with the restrictions for several years.

The way the governmental agencies handled the restrictions on sheep farms was not greatly appreciated by the people most affected: farmers. A direct source is a series of interviews conducted by a research group working under Brian Wynne, with sheep farmers from the Lake District (Cumbria, England) in 1986, including recordings of their opinions and experiences.[167] Of particular interest in these interviews is the way many farmers directly connected the events to the nearby Sellafield site and the Windscale Fire of 1957, not least because the longest bans were imposed on hill farms near Sellafield. Back in 1957, farmers in the Lake District had been told to dispose of all of their milk, despite being given almost no information on the accident and its impact. As I demonstrated in the analysis of early media reporting, the memory and unease regarding the cover-up of the Windscale Fire were directly evoked by the events of 1986. This dynamic is best illustrated in a letter to the editor published on 3 May in the *Guardian*, asking: 'Sir, am I the only reader who feels better informed about the current disaster at Chernobyl than about the 1957 fire at Windscale with which it is compared? Yours anxiously, Vince O'Connell, London'.[168] Media reporting on the sheep restrictions[169] and the community most directly impacted by these restrictions made a direct link between Chernobyl and the Sellafield-Windscale complex. Wynne's interviews show that some Lake District farmers actually thought the restrictions were the result of contamination from the Windscale Fire and the routine discharges from Sellafield. According to their reasoning, the radioactivity levels were high because they had been that high ever since 1957. The only difference between now and then, was that because of Chernobyl, action had now been taken; action that was necessary because of the British nuclear enterprise, not because of the accident in the Ukraine.[170]

Despite their affectedness by the Chernobyl fallout and the particular narrative on the sheep restrictions' link to Sellafield and Windscale, the Lake District farmers did not become distinct anti-nuclear actors in the Chernobyl debate. Wynne offers a coherent explanation:

> These more private beliefs were rarely displayed in public, and the farmers refused to confess to such dissent in media interviews. It was made clear to us that one reason for this was that the farmers identified socially with family, friends and neighbours who were part of the Sellafield industrial workforce. They recognized their own indirect and sometimes direct social dependency upon the plant – not only neighbours but also close relatives of the hill-farmers work there. Thus, underlying and bounding their expressed mistrust of the authorities and experts, there was a countervailing deep sense of social solidarity and dependency – of social identification with material kinship, friendship and community networks which needed to believe Sellafield was well-controlled and its surrounding experts credible.[171]

Thus, monetary compensation soon became the central issue in the dispute over sheep farm regulations in the Lake District. But this was also true for North Wales. Here, the triangular area between Bangor, Conwy and Dolgellau was most affected.[172] As with the Lake District, the restrictions here had led the local population to link these measures with the nearby nuclear plant in Trawsfynydd. In North Wales, the protests against the government's compensation policy went as far as farmers blockading the Welsh Office representatives' hotel during a visit to the worst affected areas.[173]

But unlike the Lake District farmers, the Welsh farmers openly communicated their views on the events to a broader audience. In January 1988, the Farmers' Union of Wales published a memorandum on the government's reaction to Chernobyl,[174] addressing it to the Agricultural Committee of the House of Commons. They then used this report as an opportunity to tell their side of the story. The fact that the farmers agreed with some aspects of the government's policy did not hide their harsh criticism of the way this policy had been implemented. Although the Union 'supported the Government's resistance to pressure to institute a slaughter programme' specifically because such a measure 'would have further undermined the public image of lamb at that time',[175] it openly expressed its resentment of the chaotic situation caused by the unpreparedness of the official institutions. These criticisms were very specific. For instance, there was not enough paint to colour-mark the sheep that had come under the restrictions. Moreover, the fact that the government had plenty of time to prepare for the situation, given the sheep slaughter occurred in late summer, did not prevent that 'by 26th August there was total confusion amongst farmers'.[176] In addition, there was much room for improvement in the payment of the compensations – until mid-December 1986, GBP 1,507,313 had been paid under the direct loss compensation scheme.[177] In order to provide the House of Commons with concrete solutions to the problems, the Union's report concluded with a list of suggestions. The most remarkable was the tenth suggestion, insofar as it reflects that the Farmers' Union of Wales had clearly identified one source of the problem in particular: 'The Government should halt the programme of nuclear power development'.[178]

The sheep farm restrictions were not only of major concern to the farmers, but also the landowners. Once again, it was a group from North Wales that communicated its views to a broader public. In 1987 the group published a report[179] openly criticizing the government's crisis management: 'The experience of 1986 appeared to the population of North Wales to be one of chaos'.[180] Like the sheep farmers, the landowners blamed MAFF for this mismanagement: 'MAFF's management throughout has been characterized by remoteness. At no time has it sought the practical advice of the farming industry and has consequently earned itself a reputation for evasion and incompetence'.[181] The report asserted that this mismanagement had led to a loss of government authority in risk definition, and as such endangered the British nuclear enterprise: 'People turned to television for their information. This is an unsatisfactory state of affairs because local opinion can be, and is, targeted wherever television programmers choose. At the moment, the target is the power station at Trawsfynydd and the government's nuclear policy. It could equally easily be shifted to any other target'.[182] The government's scientific expertise was directly scrutinized: 'Firstly, is there a need for a scientific inquiry? Secondly, how can similar incidents be better managed? These two parts are closely related because the second part implies doubt about the organizational structures at national and international levels and that implies that there is a lack of comprehension of the subject of large scale'.[183] However, the Landowners Association was less worried that the risks linked to the fallout had been underestimated. More than anything, it was concerned that the restrictions were superfluous and only aimed at covering up the government's ignorance.[184] The Association therefore called for an independent scientific inquiry, the establishment of a national institution responsible for handling situations like Chernobyl and an international indemnity fund for transborder nuclear and chemical accidents. This last recommendation highlights how the landowners, unlike the sheep farmers, identified that the greatest risks lay beyond and not within British borders. Their stance became clear when their report expressed their concern that, for instance, 'a massive accident in a French Channel reactor could cause immense damage to London'.[185]

This severe criticism of the government was further provoked by the fact that after restrictions on Welsh farms had been lifted at the end of October 1986, they were reintroduced in July 1987.[186] While the same happened in North Wales and Scotland, the restrictions were not lifted at all in the Lake District. The topic of sheep farm restrictions was prominently taken up by British newspapers in the spring and summer of 1987. They reported generally on the increasing number of farms and sheep that had come under the ban once more.[187] But they also featured the personal life stories of affected farmers and their families[188] and covered the North Wales farmers' demonstration in Whitehall, protesting against the government's compensation policies.[189] This protest was accompanied by 'a growing chorus from opposition politicians, farmers, landowners and environmental pressure groups for a full inquiry into all aspects of the disaster' – not least because in some places the records now showed radiation levels even higher than the year before.[190]

The two main environmental pressure groups that had concerns about the way the British government was handling the crisis, were Friends of the Earth (FoE) and Greenpeace. They are the only anti-nuclear power NGOs in Britain that are active nationwide. FoE is Britain's main environmental NGO. Unlike Greenpeace, it does not have a centralized organizational structure, but is an umbrella for various independent local groups; the organization has a central office in London. In addition to these two organizations, other local or regional anti-nuclear power groups have sprung up in Britain, including Cumbrians Opposed to a Radioactive Environment and the Scottish Campaign to Resist the Atomic Menace (SCRAM).[191] However, they have neither gained central importance at the national level nor succeeded in forming a publicly visible network, all the more evident since the British national anti-nuclear discourse – contrary to the French – focussed on military nuclear applications rather than on power plants.[192]

In early May 1986, FoE used Chernobyl as an opportunity to inform a wider audience about its anti-nuclear stance and, at the same time, attract new members and funds. FoE ran an advertisement on 6 May 1986 in the *Guardian* with the heading: 'The accident they said would never happen. Friends of the Earth – Campaigning against nuclear power since 1973. Join us now!'[193] The same message was communicated via various motifs. An advertisement in the

Observer a couple of days later offered a more detailed statement, which is worth citing in full:

> A positive message on Chernobyl from Friends of the Earth: It need not happen again. The nuclear accident happened 2000 miles away. Yet we in Britain were still warned not to drink fresh rainwater. What does that say about the likely affects of a similar accident here? And does anyone now doubt that it could happen? Of course we are tempted to say 'We told you so', for we warned at the Windscale Inquiry, we warned at the Sizewell Inquiry, we have warned for fifteen years that there was danger in the combination of human fallibility and technology with such unprecedented capacity for environmental harm. But there is no satisfaction in being proved right on this issue. What matters is that a similar disaster in Britain is as avoidable as it is at present predictable. Nuclear power amounts to only four per cent of Britain's energy supply. Think about it: all the danger, all that cost, all the unsolved problems of waste for just four per cent of our energy. When we have 300 years' supply of coal. When we have North Sea Oil. When we have North Sea Gas. When we haven't even started to conserve energy. When we haven't even started to explore alternative energy sources: such as sun, wind and waves. You now know that the human cost of nuclear energy is too high. This is the time to add your voice to ours. Demand that no more nuclear power stations are built. So, join us now. Or at least give us financial support. Someone has to speak on your behalf – we have the expertise and experience to do it.[194]

This text illustrates a key aspect of the British anti-nuclear power protest: since the successful campaign by British anti-nuclear activists in the late 1970s against the CEGB and government plans to build a fleet of PWRs,[195] the British anti-nuclear power protest had focussed on preventing new build projects, not on calling for a phase-out of existing plants. This was not least due to the fact that the existing British AGR design was considered less accident-prone than the imported PWRs. Another key aspect of the FoE advertisement – also typical of British anti-nuclear power arguments in general – is the question of costs. Addressed here as 'human cost', this issue has been predominantly framed in terms of purely 'financial costs': why build expensive nuclear power plants when there are cheaper alternatives available, like gas?[196]

Regarding the issue of phase-out, however, one anti-nuclear actor was clearly in favour: the Greens. This political party, like FoE, took the opportunity to communicate their position to a wider audience, for instance, in an advertisement on 8 May:

> Windscale 1957 – Three Mile Island 1977 – Chernobyl 1986 – Who really opposes nuclear power? A growing number of people. And no wonder.

> Yet ONLY ONE political party stands UNEQUIVOCALLY against nuclear power. Neither sitting on the fence, nor suffering from internal division in this crucial issue. So, which is the party to support if you oppose nuclear power? There is only one. The Green party. The Greens have ALWAYS sought a nuclear-free future. We'd stop building more nuclear plants. And start closing EXISTING ones. MAKING BRITAIN A SAFER, CLEANER PLACE. And creating thousands of jobs in conservation and renewable energy.[197]

British anti-nuclear power activists certainly did not consider the accident as just an occasion to publicize their own stance and garner more support. FoE and Greenpeace also openly challenged the official 'this cannot happen here' narrative by pointing to the insufficiencies in the containment design of British reactors.[198] Furthermore, they called for a full inquiry into the effects of the fallout in Britain. From my analysis of Chernobyl newspaper articles, their views did not play a prominent role in the media reporting. Their representatives were not attributed the role of 'counter experts'. Moreover, if FoE or Greenpeace statements were cited in articles, they were placed at the very end – not a sensible place, considering that the final paragraphs of newspaper articles are often chopped off to ensure the text fits in the page layout.

Another actor whose statements should be considered in order to understand the trajectory of the British Chernobyl debate is the Campaign for Nuclear Disarmament (CND). In 1986, the CND was the dominant force in the British anti-nuclear discourse. The campaign had been launched in 1958 and garnered extensive support for its call for unilateral nuclear disarmament. After a drop in its number of activists from the mid-1960s onwards, the CND experienced a major revival in 1979 thanks to the NATO double-track decision. Not only did hundreds of thousands of people participate in their demonstrations, they also succeeded in creating a network of activists who tracked and protested the cruise convoys in Britain – the so called Cruise Watch – and in establishing the Women's Peace Camp at the Royal Air Force base Greenham Common in Berkshire.[199] Thanks to the CND's success, the British public debate on nuclear issues in the mid-1980s revolved almost entirely around the military use of nuclear technology, with nuclear power plants playing more the role of annex to the military-dominated nuclear enterprise. CND's early communications on Chernobyl conformed to this framework in how it embedded the accident in the British disarmament debate but not put it in relation to British nuclear power plants. Two advertisements that the CND ran in British newspapers, to attract new members and donors, nicely illustrate this argument. The headline of the first was 'What if nuclear WAR was just an ACCIDENT',[200] while the other read, 'Sellafield Libya Chernobyl. "But mum, will there be a nuclear war?" You always meant to join. Fill in the coupon today'.[201] The proliferation issue was used to bridge the gap and connect their arguments to the civil nuclear enterprise as exemplified by the famous graphic depicting missiles lurking out of an open reactor, used by the CND in their campaigns after Chernobyl.[202]

Some of the very few remaining anti-nuclear power activists, who had joined forces in the 1970s for the Anti-Nuclear Campaign, hoped that Chernobyl would serve to reshape the British anti-nuclear discourse and put the issue of power plants back on the agenda. Royce Logan Turner publicly called for a relaunch of the Anti-Nuclear Campaign. His article 'Let the Campaign Roll Again' included his reasoning as to why the last campaign had ended:

> The campaign fizzled out after a short period, however, partly because the decision to go for another 10 reactors based on the PWR was quietly dropped, but mainly because it was eclipsed by CND. People decided to concentrate on stopping nuclear weapons rather than nuclear power, given the new political circumstances of the time. Cruise, and the deterioration of East–West relations, meant that nuclear bombs became a political issue, whilst nuclear power did not.[203]

However, Turner's manifesto did not manage to change the relationship between the military and civil aspects within the British anti-nuclear discourse.

The book *Something in the Wind: Politics after Chernobyl* illustrates the way Chernobyl was framed within the disarmament-dominated British anti-nuclear movement.[204] The editors, British anti-nuclear activists Louis Mackay and Mark Thompson, were both part of the European Nuclear Disarmament (END) movement.[205] In their introduction, Mackay and Thompson outlined the scope of their reflections: 'This book was germinated by Chernobyl, but [its] subject is much wider than that single catastrophe'.[206] The accident at Chernobyl was treated as a point of departure to unveil the mechanisms governing the nuclear sector in general, in the USSR, the UK and at the global level. In line with this scope, Zhores Medvedev's[207] article 'The Soviet Nuclear Energy Programme' was followed by the articles 'Accidents, Risk and Consequences' and 'Secrecy, Policy-Making and the Nuclear State'. The book's subsequent sections contain general accounts of anti-nuclear and peace movements worldwide and discuss alternative forms of electricity generation. According to the authors, the secrecy policies instated by various governments with regard to the Chernobyl fallout were a clear indication of how the nuclear system functioned in general. Further questioning the reliability of the facts that formed the basis of the accident's official evaluations, they stated 'the information provided by the Soviet Union to the 1986 IAEA symposium following the accident was generally regarded as very inadequate'.[208] The resulting IAEA and subsequent reports emphasized the statistically concealed non-existence of health effects rather than research on the long-term health effects. According to Mackay and

Thompson, the concept of radiophobia which the IAEA reports presented to explain the increase in illnesses in the most affected areas, conformed with the general communication strategies of the nuclear establishment in how it negated the real threats. *Something in the Wind: Politics after Chernobyl*, as the title suggests, aimed to look at Chernobyl from the perspective of global nuclear politics. The book did not therefore present specific statements on the health effects in Western Europe or Britain in particular. Rather, it sought to reveal the politics governing the formation of official narratives. The criticism was not only directed at the Soviet nuclear programme. Through their analysis of how Chernobyl had been handled internationally, the authors aimed to shed light on the universal structures underpinning the nuclear programme. Thus, the book's Chernobyl narrative cannot be considered anti-East European/anti-Soviet, but rather anti-nuclear in a way surpassing the East–West divide. The aspect of personal affectedness (in terms of health effects in Britain) was not addressed at all. Regarding the third category of comparison, the radiophobia-apocalypse dimension, *Something in the Wind* took a clear stance, refuting the concept of radiophobia; yet at the same time it did not elaborate on an apocalyptic scenario of the most affected regions.

Another decidedly anti-nuclear publication that situated Chernobyl within the context of global nuclear politics, is the graphic novel *Doc Chaos: The Chernobyl Effect* by Dave Thorpe.[209] Thorpe's narration of the broader context of the politics surrounding the Chernobyl accident is quite different from the political science analysis provided by Mackay and Thompson. He presents 'the real truth behind the horrors of Chernobyl'[210] in the form of a sadomasochistic science fiction story culminating in the explosion at Chernobyl. The theme is the life story of Doc Chaos: born severely disabled, this character only survived thanks to modern medicine. Having spent the greater part of his childhood in a sterile incubator, the boy – whose personality was warped by this experience – grew to become the negative ideal type of science nerd. The young man, characterized as a mean, arrogant, narcissistic, sexually obsessed, sadistic genius, graduates from Oxbridge, henceforth dedicating his life to his dream of ruling the world, exterminating and destroying everything non-scientific, and earning his money by making obscure transplants. This life story, timewise, parallels the development of nuclear energy generation: 1957, the year of the Windscale Fire, is the year when Doc Chaos learned to speak. The author interweaves his narrative of the events on the eve of the Chernobyl disaster with the Doc Chaos life story, joining the two stories at the precise moment of the explosion; the Chernobyl reactor explosion thus symbolizes the sexual intercourse Doc Chaos has with his wife Jo, the moment they create their 'own nuclear family'.

The next chapter entitled 'Don't Call It Human Error' is Thorpe's allusion to the fact that in the story, Chernobyl was not an accident but an inherent part of developing nuclear technology from its inception. *Doc Chaos: The Chernobyl Effect* was not meant to be just an apocalyptic science fiction graphic novel. The afterword by A. Lowry and N. Ramsden of the anti-nuclear World Information Service on Energy in Amsterdam, an NGO to which the proceeds of the book were donated, stated that for them, the story was grounded in reality:

> Doc Chaos is the personification of science. At least as it is practised today. A science gone out of control, accountable to no one, responsible to no one. And his/her story … is macabre, perverted, sadistic, violent, misanthropic and disgusting. But then, so is the nuclear industry. Or to be more accurate, the awful alliance between the nuclear industry and the 'destructive monster' that science has become.[211]

What is interesting about this interpretation of Chernobyl is not the actual account of the events, which are not in any case the main theme of the book. The afterword specified what the book aimed to get across: the irresponsibility of scientists towards the public. *Doc Chaos: The Chernobyl Effect* presented a clear counter narrative to the official evaluations by British and international nuclear experts, because it questioned not only their arguments and data, but the very way of doing nuclear science.

Regarding the three narrative elements, it is not possible to analyse the book in terms of personal affectedness or apocalypse/radiophobia dimensions, as the account ends when the explosions occur. The way Thorpe characterizes the plant workers, however, clearly plays with anti-East European/anti-Soviet stereotypes. Nevertheless, his criticism is not levelled at the Soviet nuclear programme. The narrative is more far-reaching than these stereotypes and may be characterized as anti-nuclear. But it goes further and even exceeds this scope, drawing on a wider anti-scientific discourse.

In France too, the statements on the impact of Chernobyl provided by public authorities and actors in the nuclear sector elicited sharp criticism. Members of the Groupement des scientifiques pour l'information sur l'énergie nucléaire (GSIEN) were the first to publicly challenge the official narrative and evaluation of the accident. GSIEN was founded in the mid-1970s by a group of physicists who opposed French nuclear policies, in particular the large-scale new build project, *Plan Messmer*. Sezin Topçu explains that the group aimed 'to bring scientific legitimacy to anti-nuclear action and to provide the movement with technical arguments'.[212] The French anti-nuclear movement was particularly strong in the 1970s, culminating in the protests against the construction of the fast breeder reactor Superphénix at Creys-Malville (Isère) in 1977. However, it neither succeeded in reversing nuclear policies nor in weakening the nuclear consensus of the French political and technical elites. The militant anti-nuclear protests of the 1970s were in many ways a sequel to the 1968 protest movement, challenging not only French nuclear policies but also the French state.[213]

In response to the early news of the accident in 1986, the GSIEN linked its interpretation of the events in Chernobyl with a general criticism of French nuclear politics.[214] But the public interest in Chernobyl soon focussed on very precise concerns: Was the radioactive fallout in France perhaps more dangerous than the authorities had claimed? Did it pose a threat to public health? Were the French nuclear authorities deliberately lying about there being no real danger? This public concern paved the way for the foundation of the first French 'counter expert' radioactivity laboratories: the Commission de recherche et d'information indépendantes sur la radioactivité (CRIIRAD) and the Association pour le contrôle de la radioactivité dans l'Ouest (ACRO). The novelty of these two organizations lay in the fact that, unlike the GSIEN members, the initiators of CRIIRAD and ACRO were not professionals from within the nuclear sector. As Sezin Topçu has aptly noted, an entirely new dynamic entered the French nuclear discourse at this time: 'Both CRIIRAD and ACRO were created by a majority of "lay-persons" (teachers, doctors, nurses, pilots, farmers, shopkeepers), aided by scientists (biologists and physicists) and technicians (ex-CEA workers, especially in the case of ACRO). Hence, unlike GSIEN in the 1970s, in the case of ACRO and CRIIRAD, scientists were no longer central to the claim for a counter competence on nuclear risks'.[215] In the years to come, CRIIRAD and its founding director Michèle Rivasi would play a leading role in the French Chernobyl debate, dedicating themselves to keeping alive the public memory of the *affaire Tchernobyl*. This actor's role is easier to describe and understand once it is clear how the *affaire Tchernobyl* emerged.[216]

In the days immediately following the accident, statements released by SCPRI, IPSN and EDF representatives dominated the French media reports on Chernobyl. Journalists had turned to these authorities to obtain information on the impact of the accident in Eastern and Western Europe. The explanatory pattern used by these agencies – to criticize other countries for their overreaction and at the same time stress that there was no reason to worry about the radioactive fallout in France – had been conveyed directly to the French public. However, by the end of the first week in May, progressively more journalists were beginning to wonder why protective measures had been instated and even intensified on the other side of the Rhine in Germany, while those they interviewed in France continued to insist that everything was fine. In an article published on 9 May 1986 in *Libération*, the author speculated that 'it would, however, be extraordinary if Alsace and Lorraine, like a paradisaical island, totally escape this fallout that is so important just some kilometres further away'.[217] At this time more journalists began criticizing how the SCPRI and the nuclear sector provided them with information. One journalist, Hélène Crié, writing for *Libération*, was particularly blunt with her criticism, embedding it in a general attack on the French nuclear sector:

> The most nuclearized country in the world, France, is also the most avaricious with regard to information. Even when the accidents do not take place on her soil. … However, it is not the competent and official authorities that are lacking. But, as soon as it comes to obtaining the slightest bit of information, you run up against a brick wall. … With regard to nuclear, three partners communicate with each other, but only with each other: the constructors, the operators and the public authorities. … The only experts that are capable of providing information on the risks, belong to the family that produces this type of energy.[218]

To alleviate this growing criticism of French nuclear policies, the public authorities adopted a strategy of providing clear counter statements. The press communiqué released by the French Minister of Agriculture on 6 May exemplifies this strategy: 'The French territory, because of its distance, was completely exempted from the fallout of the radionuclides consequent to the accident at the Chernobyl plant'.[219] However, these unrealistic statements only instigated more public speculations on the 'true impact' of Chernobyl in France. To put an end to these speculations, Pierre Pellerin of the SCPRI took part in a TV interview on 10 May. He appeared on the show *13 h* on channel TF1 and presented maps indicating the levels of radioactivity from the Chernobyl fallout over continental France.[220] Monique Sené, the president of GSIEN, had been invited to speak on the same show. At the start of the broadcast, she had a heated argument with Pellerin about the protective measures implemented in Italy and West Germany. She claimed the measures were justified, a statement Pellerin did not agree with at all. Furthermore, Sené then accused Pellerin of having withheld information; an accusation Pellerin refuted by highlighting the more than 200 Telex messages the SCPRI had sent to journalists, politicians and others since news of the accident had been communicated to them. After this argument, the talk show hosts asked Pellerin to present the maps showing the radioactivity the SCPRI had measured in France in late April and early May. Pellerin obliged: he showed one for every single day, from April 28 to May 5. Then followed another map indicating the total deposition for each region. All the data was measured in the unit Curie, which Monique Sené directly criticized as an attempt to make the data evaluation more complicated, seeing as the Becquerel had become a more standard unit of measurement. While Pellerin presented his maps, the show host interrupted him to ask how far these levels were above the normal background radiation. When Pellerin replied that the radioactivity had been up to 400 times higher than normal in some local spots, the interviewer exclaimed: 'These are numbers that we discover today!' But as with his first Chernobyl interview on 29 April, Pellerin paid very little attention to the interruptions of others and continued his presentation. However, 'These are numbers that we discover today!' was the statement people remembered about this interview, which can be considered the starting point of the *affaire Tchernobyl*. After the TV interview, many people concluded that over the previous fortnight, the radiation protection experts and the government had lied to the French public and covered up the true amount of radioactive fallout that had hit France. And in doing so, these authorities had put French public health at risk. Some journalists felt personally betrayed, especially since they were the ones who had passed the message to their audience that there was no need for any protective measures. So, what would happen if the reactions and measures taken in other countries indeed proved to be appropriate, while in France the decision to maintain the status quo had placed the French people's health at risk?[221]

On 11 May, the day after the interview with Pierre Pellerin had been broadcast, the French Minister for the Environment, Alain Carignon, was confronted in a news programme on channel Antenne 2 with the accusation that the state authorities had purposefully withheld important information. The programme's host began the interview by asking: 'How do you explain the fact that the French have been so badly informed in this affair, because we journalists, we tried to obtain information on these issues and we were given the wrong information'.[222] Carignon refuted this allegation and declared that the data had all been available before, the only new element was that the SCPRI had gathered the data onto one single map, removing the background radiation. However, the news host insisted that this data had not been available to the public, and asked Carignon: 'How do you explain that the French never knew about them?' Carignon parried this question with the reply: 'I beg your pardon, but it is the task of the media to collect these numbers'. The question about who had access to what data was not resolved in the interview, nor has it been even to this day. Instead, a *cellule de coordination de l'information* was created to prevent further misunderstandings between the French nuclear experts and journalists in the event of a nuclear accident. The French authorities had identified their communication with the public as the main area requiring improvement. After all, Pierre Pellerin and his colleagues were scientists, not public relations managers; had there been better communication with the media, the whole *affaire Tchernobyl* would never have happened. State officials, however, never questioned their stance on the health risks linked to the radioactive fallout. Carignon's declaration in the interview that 'these amounts are not dangerous' conformed perfectly to this standpoint.

The fact that the French government and nuclear authorities saw no reason to re-evaluate their position that the fallout in France was not dangerous and that protective measures were unnecessary, however, did not mean that everybody else felt the same. On Monday, 12 May, after Pellerin and Carignon had made their statements on TV that weekend, newsstands all over the country displayed the headline on the cover of *Libération*: 'The Radioactive Lie' with this statement underneath: 'The public authorities in France lied, the radioactive cloud from Chernobyl did indeed fly over a part of the Hexagon: Professor Pellerin in fact admitted it two weeks after the nuclear accident'.[223] On TV, the news anchors repeatedly stated that they had not received the correct information from the public authorities. A news bulletin on channel Antenne 2 informed its audience that: 'Between 1 and 4 May, not even a single official communiqué. On these dates, the cloud advances and swells over France. … With no particular information, no media reports on this cloud'.[224] After 4 May, however, Pierre Pellerin provided the information that the cloud had moved on, but nothing had been said about the levels of radioactivity on the ground: 'Reassuring comments, but we have not been given any data'. Thus, the *mensonge de l'état* (the state lie) narrative of Chernobyl was born. Moreover, the person responsible for putting the French public at risk had been directly identified: Pierre Pellerin. The fact that Pellerin had not come across as a particularly nice and empathetic person in the TV interviews, but appeared to be arrogant and gruff, helped to portray him as 'the bad *nucléocrate*'.[225] Media reports were replete with accusations against Pellerin and speculations as to who was involved in this French state plot against its own people.[226] One possible motive for this policy of secrecy soon surfaced: the government had wanted to prevent the French agricultural sector from being hit with bad publicity: rumours about radioactive contamination of French foodstuffs could have had disastrous consequences for its export sector. The regional newspaper *Les Dernières Nouvelles d'Alsace* wrote on 12 May, 'the consumer might get the feeling that the interests of the French farmers have been set above the general interest'.[227] But the journalists did not only blame certain individuals for the wrong decisions, they also put the *affaire Tchernobyl* in the wider context of the French nuclear sector – its structures and how it functioned – and its entanglement with the French state. Thus, they used Chernobyl to criticize the French nuclear techno-political regime as a whole. The notions of 'disinformation' and 'secrecy' in particular drew heavily on the 'nucleocratic' frame, which constituted a central topos of French anti-nuclear criticism. A particularly telling example is in an article published on 13 May in *Libération*:

> Since the start of the adventure in 1956, France has placed herself resolutely in the heart of secrecy. Sometimes even escaping the politicians, the atom is the privilege of a few great engineers who graduated from X [short for Ecole Polytechnique] and mines [short for Ecole des Mines]. … From the beginning, the 'X-mines' had a hold on the atom, which turned into a family affair, i.e. an affair of corps and clan. Pierre Guillaumat [[228]], their president, always carried

> a little notebook with the names of the corpsards (ingénieurs des mines). He himself placed them in the state apparatus, applying unwritten but terribly subtle laws.[229]

It was without doubt primarily the left-wing press articles that framed the *affaire Tchernobyl* in an anti-state manner. Another example is the article of 12 May in *Rouge*, a Jeunesse communiste révolutionnaire magazine, that described the group responsible for the secrecy surrounding Chernobyl's impact in France as a 'cast of conspirators graduated from the Grandes écoles, Polytechnique, les Mines etc., the masters of nuclear'.[230] The very strong anti-state sentiment that characterized the French anti-nuclear movement of the 1970s was promptly revived during the *affaire Tchernobyl*. Furthermore, this stance found in Chernobyl an outlet to express itself after the French anti-nuclear movement had lost momentum by the late 1970s.

The left-leaning media, however, were not alone in criticizing the French government and the nuclear authorities for how they had handled Chernobyl. The lack of independent expertise in the French nuclear sector was also addressed in an article in *L'Express*: 'What will change with this new unit put in place by Alain Madelin? Precisely nothing. Not a single independent scientist, not one syndicalist, not one ecologist, not one specialized journalist will be part of it'.[231] This gap would ultimately be filled with the new 'counter expertise' that had emerged from the *affaire Tchernobyl*: CRIIRAD and ACRO. For this reason, their statements gained prominence in the development of the French Chernobyl debate. Furthermore, not quick to forget how in their opinion they had been misled by state experts, journalists no longer relied exclusively on the national radiation protection agencies for their information, but now also turned to these 'counter experts' when reporting on radiation issues.

Individual Voices

This section is dedicated exclusively to British publications. As far as I could discern, only a handful of books on Chernobyl were published in France in the two years following the accident. Of these, the books *Tchernobyl-sur-Seine* by Hélène Crié and Yves Lenoir, and *L'affaire Tchernobyl* by Yves Lecerf and Edouard Parker likely reached the broadest audience.[232] Naming this section 'Individual Voices' does not imply that individual agency did not play a role with regard to the actors analysed in the previous sections. Lord Marshall of Goring, Pierre Pellerin, Monique Sené and many others acted as individuals in the Chernobyl debate. Driven by their personal convictions, they considered it important to convey their evaluation of Chernobyl to a broader public. Yet, these people were perceived as closely associated with a certain institution or organization: Lord Marshall of Goring with the CEGB, Pierre Pellerin with the SCPRI and Monique Sené with GSIEN. The actors in the Chernobyl debate who I discuss in this section do not represent or act as spokesperson for a particular institution. Moreover, unlike Louis Mackay and Mark Thompson, or Dave Thorpe, they do not ally with particular institutions in their publications. For all these reasons, this actor cluster has been named 'Individual Voices'. Interestingly, these individual voices did indeed exist in Britain, whereas in France, the debate was directly taken over by the established actors of the general nuclear debate. The early Chernobyl debate in Britain was thus more open than in France in the way it integrated a broader spectrum of actors.

One such individual voice in the British Chernobyl debate is Richard F. Mould. He published his first book on Chernobyl in 1988. For this popular science publication, Mould chose the title *Chernobyl: The Real Story*.[233] Mould presented himself to the reader as a neutral expert, capable of delivering a trustworthy judgement of what really happened. Leaving no doubts as to his intentions, the opening sentence of his preface read: 'Now, what I want is Facts'[234] and clarified: 'My objective in writing this book was to produce an historical account of what happened before, during and after the accident, rather than embroiling myself in the political arena over Chernobyl's implication for the environmental issue'.[235] In order to prove his statements, Mould included 'as much photographic evidence as possible' in his book – starting with photos of himself visiting the Chernobyl plant in 1987. Mould considered his authority as a neutral expert was derived from his profession and therefore described himself as

> an internationally known medical physicist, cancer statistician and radiation historian. As well as serving as a technical expert to the World Health Organisation and the International Atomic Energy Agency, [I] was a United Kingdom Government delegate to the IAEA Chernobyl Post-Accident Review meeting in Vienna, 25–29 August 1986.[236]

This statement was apparently convincing, because in the following years, Mould's book featured in numerous British Chernobyl bibliographies, including research publications that listed only natural science papers and reports. This is why I devote so much space to analysing the narrative of this popular work.

The first few pages of the book are a technical description of the Chernobyl nuclear power plant; several paragraphs on Kiev underline the fact that the town had luckily been saved from the worst of the fallout and that the river Dnieper 'was not contaminated'; and a minute-by-minute account of how the accident unfolded. The series of photos inserted after this section provide a particular narrative. The first photo shows the lively and well kept main street of Kiev in 1986, whereas the picture below shows this street in 1943 after its destruction during Second World War, with the corresponding caption: 'This is not a nuclear landscape, but the aftermath of the German invasion of the Ukraine during World War II'.[237] The next page shows a smiling bridal couple in the streets of Kiev in May 1986 – no caption included here, and therefore no information imparted to the reader regarding the fact that in May 1986, the people of Kiev had not been warned or told about the high levels of radioactivity in the air. By letting these pictures speak for themselves, as the author intended, they told the story of a town unaffected by the Chernobyl fallout, but which had survived disasters on an incomparable scale to this accident, disasters that had nothing to do with radioactivity but with German belligerence. Throughout the chapter, there are more photographs that – through the interplay of captions or no captions – imply a particular narrative. One photograph captures a group of workers as they were preparing for their shift on the reactor roof. The caption reads: 'The protective suits are constructed of lead and rubber'.[238] The important additional information, that these suits were often taken off because workers could barely move in them, was not mentioned in the caption, nor in the main text. Mould did not, however, ignore the devastating side of the accident, illustrated with photos of firemen brought to a specialized hospital in Moscow, revealing them lying in bed, bald and with black marks all over their bodies.[239] Yet, the imagery did not form a coherent narrative and was completed with peculiar pictures such as one of a bottle of Ukrainian vodka with the brief description: 'Original firewater prepared according to selected traditional Ukrainian recipes forms a spirit of the highest quality with the addition of natural honey and blackcurrants. Refrigeration is recommended before use'.[240] Mould left it to the reader to decide how this item should be included in the accident's narrative.

The next chapter, 'The Victims', begins with a broad definition of the term 'Chernobyl victim': Mould elaborated that not only the thirty-one direct fatalities had fallen prey to the accident, but that 'the 135,000 evacuees from the 30-kilometer zone and the people who eventually will be included in the number of excess cancer deaths statistics are also victims of the disaster, as I suspect will be some national nuclear power plant programmes for the production of electricity'.[241] Severe health effects resulting from the radiation exposure were presented as being limited to those who had been at the plant site: '299 were diagnosed as having radiation syndrome, but these cases were confined to firemen and plant workers and there were none in the general population'. As for the concerns about other kinds of health effects, Mould treated them as though they were unfounded or even ridiculous:

> There were recorded cases of people diagnosing themselves as having radiation sickness when all they had was a stomach upset. More unusual stories

> included the vegetarian who survived 3 days by eating only peanuts, the lady who wanted to know if she should dodge rainspots because of radioactive fallout, and the newly returned visitor from Eastern Europe who wanted to know if he was radioactive because an East European lady had recently been breathing heavily on him.[242]

An anecdote about the IAEA conference in Vienna in August 1986, which Mould had attended, also conformed to his viewpoint: 'Some Soviet officials were against the idea of a visit from Dr Hans Blix, Director General of the IAEA, and gave as their reason that they "were worried that the radiation might harm Dr Blix's organism."'[243] Apparently, this exaggerated fear had raised 'a certain amount of laughter'[244] among the assembled group of Western radiation experts. Mould's closing remark to this chapter that 'not all of the irradiated firemen victims died'[245] was designed to add yet another example to his argument that the health effects of the accident were not as devastating as many people believed. To come up with such exaggerated ideas, people simply must have been influenced too much by the media reports which had seminated panic. Therefore, 'one of the resolutions following the accident should be to educate the general public, as far as possible, on radiation risks and benefits (the benefits of radiation treatment and diagnosis in medicine should not be forgotten)'.[246] The photos illustrating this chapter underpinned Mould's narrative that the health effects were limited. The series of pictures started with a portrait of 'one of the oldest evacuees', smiling happily, like the others in the photo. The same happiness and easiness were visible in the next photograph, too, which depicted evacuated children at 'a summer camp'.[247] The subject of happy, healthy evacuees was revisited at the end of the photo series in a picture showing a group of smiling people inaugurating a new settlement for the clean-up workers.[248] Of note here are the pictures of the firemen who survived the accident: a group of smiling middle-aged men standing in the sun outside a rehabilitation centre. The caption explained why some of the men were wearing hats: 'The caps cover bald scalps which have been shaved as part of the decontamination process'.[249] This is indeed an interesting explanation considering these men had just endured a radiation sickness that might induced severe hairloss. No less interesting is the caption alongside the next picture depicting three of the Chernobyl firemen on exercise machines: 'Lung capacity of the firemen could well have been impaired due to inhalation of smoke, dust and dirt'.[250] Not a single word was mentioned about the possible health effects of the radionuclides they had inhaled. To what extent Mould himself believed these photographs, or wanted to believe in them, can be deduced from his comment in the chapter on the evacuations. A 'photograph [that] showed seven of the babies all well wrapped up from head to toe, like Russian dolls, with only their little faces peeping out!' was for him proof that these babies born to evacuated mothers were all doing well.[251]

Of the other chapters addressing 'The Decontamination of the Environment', 'The Entombment of the Reactor' and 'The Food Chain', it is interesting to examine the last one. Mould lists the radiation levels observed and the countermeasures taken in various European countries. His remarks with regard to the UK are brief and include the statement: 'Levels in milk are below the recommended levels at which restrictions on milk supplies would be considered in the United Kingdom'.[252] While this was not inaccurate information, it is interesting that Mould fails to mention even once the radiation levels that had been registered in sheep meat. After all, at the moment of publication, hundreds of sheep farms in the UK were still subject to the restrictions put into place shortly after the Chernobyl accident because the animals were still too radioactive to be sent to market.

Like the authors of the Watt Committee report, Mould considered the Chernobyl accident above all as an interesting subject for natural science research. To his account on the babies born to evacuated mothers he added: 'It is to be hoped that all these babies will receive medical follow-ups for the rest of their lives. This will ensure, not only that they will promptly receive any necessary treatment, but it will also provide extremely valuable data for the future on the effects of low-level radiation doses'.[253] Although Mould stressed there were still many unknowns regarding the long-term health effects of Chernobyl, he considered the death toll estimate in an IAEA report dating from August 1986 – written by Dan Beninson, Director for Licensing Nuclear Installations in Argentina – to be the most trustworthy and referred to it: Beninson had calculated 2,000 excess deaths and related this number directly to the health impact of naturally occurring background radiation.[254]

In summary, the narrative Mould presented in *Chernobyl: The Real Story*, emphasizes the viewpoint that people's fear of radiation and the possible health effects were exaggerated. The implicit belief was quite clearly conveyed that though the Chernobyl accident had caused many deaths – each one was one too many – people had to learn to put these deaths better into perspective, for the simple reason that: 'like it or not, civil nuclear power remains the only viable alternative for the foreseeable future'.[255] The fact that this statement did not appear until the end of the book is misleading. Mould did not actually deduce this conclusion from the information provided. The facts he presented were part of the underlying assumption that, in the end, the impact of the Chernobyl accident could not have been so terrible nor so terribly shocking because we agreed to live with its consequences when we entered the nuclear age.

Once we analyse Mould's narrative for the presence and weight assigned to the three main components, it is obvious that personal affectedness does not play a role. This is particularly evident given the fact that Mould did not deem it important to mention the restrictions on British sheep farms. Anti-East European/anti-Soviet stereotypes do feature in his narrative: the photograph of the vodka bottle is a telling example. In addition, the way Mould portrayed the evacuees depicted them as a strong natured but simple (minded), rural and traditional people who willingly did what they were told. Mould played on this stereotype to prove his point that the health impact of the accident was limited: the little Russian children and baby 'dolls' looked so healthy and prosperous, so what could possibly be wrong with them? In general, Mould's account of the health effects was far from apocalyptic. The apocalypse for Ukraine was not Chernobyl; it had been the German occupation during Second World War. When we compare the ruins of Kiev in the aftermath of Second World War, with the images of Kiev and the new towns built for the evacuees following the Chernobyl accident, the smart streets are full of happy smiling faces in the latter images. Like the British authorities and nuclear power industry actors, Mould identified the need to better educate the public about the true scale of the risks of ionizing radiation and its benefits as the main conclusion from the Chernobyl accident. At no point did he question the nuclear programme as such, and he actually considered nuclear power production a fact of life to which people simply had to adapt. Mould's self-promotion as a neutral nuclear expert was willingly taken up by many people who shared his convictions about Chernobyl. Referring to Mould's book created the illusion of using an independent, neutral source that was not informed by an inherent pro-nuclear point of view.

Other individual voices questioned the state and nuclear sector's official narratives and shed light on the dynamics and power structures at play in the formulation of these evaluations. One of the earliest of these counter narratives was the book *Mayday at Chernobyl* by Henry Hamman and Stuart Parrott.[256] A Radio Free Europe journalist based in London, and an East European history and politics specialist, respectively, they presented a critical assessment of the reports published by the official institutions. From the outset, these two authors made their opinions known regarding the various governments' emergency measures and stated that politicians and nuclear power supporters had insisted that there was no real danger while the fallout was already spreading across Europe. The aim of their publication, however, was not to offer a concrete counter evaluation of this fallout, but to cast light on the context generating the international nuclear community's narratives on Chernobyl. The authors' often sarcastic approach was an attempt to direct attention to the implicit immorality of many of these official statements. A telling example is their comment on what so many radiation scientists had professed an interest in, namely the 'positive learning opportunity' that Chernobyl had given them: 'The explosion of the American atom bombs at Hiroshima and Nagasaki provided scientists with a unique opportunity to study the devastating consequences of exposure to radiation and the effects of radiation sickness. Now, the accident at Chernobyl will also provide unique opportunities for study and research'.[257] To give readers an overview of the aspects relevant to the Chernobyl debate, the book discussed the nuclear history of the USSR; how the accident had been covered up by the Soviet government and media; the dispersion of the radioactive plume; and the widespread fallout. Moreover, an entire chapter was dedicated to the 'liquidators'. As for the data available on Chernobyl's health impact, Hamman and Parrott included a wide selection of sources and duly noted any areas where knowledge was still vague or patchy. In their detailed account of the low-level radiation debate and the different points of views, rather than presenting their 'truth', they concluded that 'the effect of low-level radiation exposure is one of the most politically charged issues in science'.[258] The authors applied the same caution when communicating information on the death toll: they presented different estimates but no final judgement on these figures.

Despite their balanced account of the fallout's health impact, Hamman and Parrott take a clear stance on how states carried out their risk management. This is especially true for the chapter 'No One Thought It Could Happen'. Their account of the events, for the Soviet Union as well as for Western Europe, stands in direct contrast to narratives by British authorities: 'The nuclear accident caught European governments by surprise, and their response was slow and confused. The problem everywhere was inadequate planning'.[259] The unpreparedness of governments is what led to their adoption of tactics and strategies to cover up this incompetence; according to Hamman and Parrott, this was particularly true for Belgium and France.

The authors did not limit their criticism to individual national governments. They analyse this action, or non-action, within the bigger framework of nuclear politics cultivated by the IAEA. In their chapter 'Atomic Politics' they convey a detailed interpretation. According to Hamman and Parrott, the IAEA was 'not meant to be a neutral inspectorate or regulatory board but a nuclear energy promoter and a forum for nuclear power vendors to display the plants they have to sell'.[260] In the immediate aftermath of the accident, the IAEA's task was nothing less than to save the nuclear industry as a whole. To this end, intensive behind-closed-doors politics were undertaken during the summer of 1986 to find an official narrative that suited everybody. Hammam and Parrott's argument is worth quoting in detail:

> The crucial battle was the one between the Soviet Union and the Western nuclear powers. The struggle was over the question of placing the blame for the Chernobyl accident. The Western nuclear strategy was clear from early on – to argue that the Soviet nuclear programme was both different from and inferior to Western programmes. In the halls around the IAEA board room, the word went out from the French, the West Germans, the British

> and the Americans that the RBMK was a badly designed reactor that lacked containment and could not be licensed in the West. A French delegate said that the goal in his mind, at least, was to make sure that when people looked at the Chernobyl accident, they drew the 'correct conclusions'. The Soviet side preferred to see Chernobyl as one of a series of major technical disasters. They pointed to the American Three Mile Island accident, the American Challenger spacecraft explosion and the Union Carbide disaster in Bhopal. What linked them all, in the Soviet view, was that man had failed to deal adequately with technology. For the Soviets, the goal was to present Chernobyl as a human failure rather than a failure of Soviet design. This was not what the West wanted for a verdict. If human fallibility was predominantly the cause of Chernobyl, then how could Western plants be safe from similar human failures? Throughout the summer, the battle was fought out sotto voce. Finally, a compromise was reached: the failure was to be attributed to the 'man-machine-interface', the way that the operators interacted with the technology. It was a phrase both sides could live with.[261]

Thus, Hamman and Parrott's criticism was less about concrete numbers, than about the politics that had dictated and conditioned the establishment of the official facts presented to the public as absolute truths not as the lowest common denominator between clashing economic interests. Like Mackay and Thompson's publication, *Mayday at Chernobyl* focussed on exposing the universal structures of the global nuclear programme through the lens of the Chernobyl accident. Instead of providing a coherent narrative on the causes, evolution and consequences of the accident, Hamman and Parrott pointed out the still unanswered questions and reflected upon the politics underlying the creation of scientific facts in nuclear energy.

In 1988, two other authors chose a similar approach for their account *The Chernobyl Disaster*.[262] Viktor Haynes and Marko Bojcun, academics working in the field of Eastern European studies, use their book's subtitle to summarize their interpretation of the events: 'The True Story of a Catastrophe – an Unanswerable Indictment of Nuclear Power'. Based on a wide range of sources, including eyewitness reports, Russian and Ukrainian newspaper articles, and IAEA papers, the authors give a detailed description of RBMK technology, the progression of the accident as it unfolded, the spread of the radioactive plume, the evacuations and the estimates of the fallout's health effects. Regarding the Soviet and British radiation experts' estimates, Haynes and Bojcun were particularly shocked at the systematic comparison between the Chernobyl death toll and the overall number of cancer deaths which, to use their words, 'strikes one as perverse'.[263] 'The logic of such an argument leads inexorably to the view that nuclear accidents don't matter, as they affect only thousands of people while millions will die anyway; and from there to the notion that killing people is justified because they will die eventually in any case'.[264] Moreover, the authors felt that comparing the fallout intensity with natural background radiation was particularly dishonest, insofar as the natural background radiation was largely a man-made phenomenon caused by the 'routine and accidental emissions from nuclear power stations, uranium mining and atomic blasts, as well as by medical treatment involving radioactivity'.[265] Haynes and Bojcun did not present a concrete death toll, but pointed to the implicit assumptions underpinning the various calculations and spoke of a range 'between a few thousands to over 100,000 people (depending on which experts are to be-lieved) over the next half a century'.[266] In doing so, they made a very important assumption themselves: 'As long as it was thought that low-level radiation was not dangerous, nuclear reactors were deemed to be "safe". Evidence has shown that any amount of radiation is dangerous'.[267] Many radiation protection specialists would not agree at all with this statement.

When examining the question of who was to blame, the authors presented a response that directly opposed the official narrative by the IAEA and British nuclear experts. According to Haynes and Bojcun, it was unreasonable to saddle workers with the responsibility for the accident. It ought to have been allocated at a much higher level in the hierarchy: 'The faults in construction of Chernobyl's Number 4 reactor, the difficulties in its operation and the lack of a containment structure to withstand the explosion cannot be blamed on its builders and the station's operators and managers. The responsibility rests with the Politburo, Gosplan and the ministries in charge of construction'.[268] Haynes and Bojcun also went on to contest the narrative of the well organized evacuation. They dedicated a whole chapter to 'Unnecessary Irradiation' and concluded that the commission responsible for the evacuations had 'failed miserably to protect the health not only of those living within the five kilometres of the station',[269] but also those in a more extensive geographic area. Criticism of the Soviet political system is evident throughout the book, but it is perhaps most clearly and explicitly formulated in the chapter 'Lessons of Chernobyl': 'Simply put, the disaster assumed its full dimensions as the result of a particular system of political rule, of the habits and self-interest of its bureaucratic elite, of the ways in which it has become accustomed to valuing the security, prestige and economic might of the state above the welfare of its labouring classes'.[270] Haynes and Bojcun did not stop their criticism here, however, and went on to write: 'We do not suggest for one moment that the leaders of Western nuclear power states behave in a fundamentally different way. The participation of the IAEA member states in the cover-up of the Chernobyl disaster speaks for itself'.[271] Thus, the authors believed the problem was on an entirely different scale, surpassing the boundaries of every political system: it was to be found in the logic inherent to industrial societies and their thirst for power and constant growth.

Regarding the aftermath of Chernobyl, the book *Britain's Nuclear Nightmare* by James Cutler and Rob Edwards is probably the most important example of how British anti-nuclear power activists placed Chernobyl in relation to the British nuclear enterprise.[272] The book's main topic is the Sellafield-Windscale complex. It provides detailed descriptions of the history of the site, the Windscale Fire, the discharges released into the Irish Sea and the cancer rates in children living near the site. In addition, it addresses issues associated with the working conditions of plant workers and proliferation issues. In chapter 9, 'Patterns of Deceit', the authors incorporate Chernobyl into their considerations. The chapter opens with:

> Three days after Chernobyl exploded, Britain's Environment Secretary, Kenneth Baker, told the House of Commons: 'There is openness and frankness in this country in dealing with the nuclear industry … [*sic*] if there had been an accident of that kind in this country, there is no question, but that we would have been open and frank about it straight away.' Britain's honesty, the Government suggested, contrasted favourably with the Soviet Union's secretiveness. Yet the full report of the inquiry into Britain's only comparable accident at Windscale was only published three decades after the event.[273]

Yet, this blatant secretiveness had not begun with Windscale, it was an intrinsic feature of the British nuclear enterprise from its inception. To prove this point, Cutler and Edwards refer to the main authority on British nuclear history, Margaret Gowing.[274]

> According to the official atomic historian, Professor Margaret Gowing, Britain's post-war Prime Minister, Clement Atlee, was 'obsessively secretive' about the development of the atomic bomb. In a lecture at Cambridge in 1978, she highlighted the secrecy in which all the early decision-making was shrouded: 'The Cabinet as a body was completely excluded from major atomic decision-making. A small inner ring of senior ministers took decisions in a confusing number of ad hoc committees with science fiction titles which never reported to the Cabinet.'[275]

This built-in secretiveness did not just apply to the military applications of this technology. The same was also true for electricity generation: 'When it came to beginning the development of nuclear power for civil purposes, even the chairman of the electricity generating authority, Sir Walter Citrine, was kept in the dark. In 1954, when the Government convened a working party to discuss plans for Britain's first reactor programme, it never told Sir Walter of its existence.'[276] And this secretiveness was very much preserved even during the 1980s: 'When the Government papers for 1954 were released in 1984, a large number relating to nuclear energy were withheld, some of them under a provision which enabled them to be kept under wraps forever'.[277] The change that Chernobyl brought to this setting was 'a veritable avalanche of brave new intentions'.[278] However, in the authors' view, these initiatives were primarily superficial. One example in 1986, was when Sellafield's chairman Christopher Harding 'launched a £2 million advertising campaign to try to close the "comprehension gap" which had been opened up on the subject of nuclear power'.[279]

In the final chapter, 'The Faustian Bargain', the authors again refer to Chernobyl to reinforce their argument. This chapter opens with a fictional description of an accident at Sellafield. This scenario was intended to directly challenge the British 'industry's most persistent post-Chernobyl riposte: "It could not happen here."'[280] The authors argued that such a statement needed to be clarified, specifying: 'Britain does not run any reactors designed exactly like the Soviet reactor, so a Chernobyl-type accident is logically impossible. A Chernobyl-scale accident, though, is quite possible'.[281]

Cutler and Edwards spoke of Chernobyl as more than just an abstract event. They provided information about its impact on public health, and rather than adhere to one estimate or another, they supplied the two extremes of the death toll estimates – and by doing so, drew attention to their dramatic disparity: 'According to official estimates it will kill up to 2,000 people in Europe, including perhaps forty-five in Britain, and as many as 40,000 people in the Soviet Union. According to Ernst Sternglass, radiology professor at the American University of Pittsburgh, the eventual overall death toll could exceed 1,000,000'.[282] To illustrate the global impact of the fallout, the authors pointed to the case of Britain: 'Eighteen months later, more than 700 farms in Cumbria, Wales, Scotland and Northern Ireland were still affected by restrictions on sheep movement and slaughter because of radiation from the accident'.[283] From here they directly identified a grave error and failure on the part of the British nuclear authorities: 'The response of the monitoring and regulatory agencies in Britain was manifestly incompetent. Dozens of children throughout the country will contract thyroid cancer over the next thirty years because of the Government's failure to warn of the dangers of drinking Chernobyl-contaminated milk'.[284] The greatest failure, however, was that the British government had not learnt the right lesson from the accident in terms of its energy policies. While almost all European countries had never embarked on the nuclear enterprise, or had abandoned their plans, or now, after Chernobyl, at least postponed their new build projects, 'there were only three exceptions to the rule: the USSR, France and Britain'.[285] The authors ended their account with a positive message to the British public:

> In the last resort, the decisions which have to be taken about nuclear power and nuclear weapons are not technical or economic, they are political. The

> only necessity for the abandonment of civil and military nuclear technology is political will. The essential prerequisite for public will is public demand. Unlike poor Dr Faustus, the British public still has a choice. Even at the eleventh hour, we can escape damnation. We can opt for a future powered and perverted by plutonium, or a future free from nuclear contamination. We can choose to wake from Britain's nuclear nightmare.[286]

Three reasons underpinned my decision to quote this book in such detail. Firstly, by highlighting the authors' criticisms, I could show that similar language was used to describe the British nuclear sector and the French nuclear sector, namely with regard to the secrecy with which it carried out its affairs and its description as a sphere restricted to a close circle of insiders. Secondly, these quotes aptly illustrate the close connection Britain made between Chernobyl and the Sellafield-Windscale complex. As I have shown above, this link was made in media reports and by local sheep farmers. *Britain's Nuclear Nightmare* furthermore reveals the importance of the Sellafield-Windscale complex in the British anti-nuclear argument. Even in the direct aftermath of Chernobyl, the main concern was not how many people could possibly die in Britain from the fallout, but rather the risk Sellafield presented. Chernobyl's impact in Britain figured more as a footnote in *Britain's Nuclear Nightmare* to illustrate the geographical reach of a large-scale nuclear accident – and the risk that the next large-scale nuclear accident could soon occur was right there in Britain, crouched on the Cumbrian coast at Sellafield. Thirdly, the quotes make it blatantly obvious that even an entirely anti-power-plant-argument could not be considered in Britain without linking the issue to the military nuclear complex.

Conclusion

A comparison of the initial statements released by French and British authorities in response to news reports about Chernobyl demonstrates their similarity.[287] In both countries, state officials focussed on dismissing rumours that the fallout from Chernobyl could have any negative consequences for their countries.[288] In addition, they stressed that the national plants were simply not comparable to Chernobyl because the reactor designs in the West were completely different from those used in the USSR. Thus, they declared there was no cause for concern, either about the fallout or the risk that a similar accident could happen in the homeland. However, in France, by the second week in May, the very same authorities were publicly confronted in the media – accused of deliberately holding back the real data on radioactive fallout in France in order to protect the agricultural sector and the French nuclear enterprise from negative publicity. These accusations were put forward by anti-nuclear activists and alarmed journalists who had closely observed the protective measures taken in other Western European countries, particularly West Germany and Switzerland. These critical voices were puzzled about the disparity between the measures taken abroad and the lack of official concern in France; they therefore questioned the statement that France had not been affected by any serious fallout. This dispute came to be commonly referred to as the *affaire Tchernobyl*. The counter side expressed their criticism of the official French response to Chernobyl and the attitude of the state experts, traditionally adopting the French anti-nuclear strategy with strong anti-state arguments. The *nucléocratie* and the internal logic of the entire French nuclear techno-political regime were blamed for the 'lies' and 'cover-ups' surrounding the Chernobyl accident and how it had been handled thus far. The French Chernobyl debate very quickly reached national proportions and focussed on the role of the French state experts. Precisely this role, which these nuclear experts played within the power structures of the French state, was called into question in the *affaire Tchernobyl*.

British discussions in the immediate aftermath of Chernobyl were quite different, although it soon came to light that the public authorities had completely misjudged the fallout's impact in the UK. Their predictions were proven wrong, not once but several times. Firstly, in early May 1986, rainwater and milk were found to exceed radiation safety levels in some parts of Britain. Secondly, in June 1986, upland sheep farmers were informed that they could not continue to handle their animals as usual; over the previous weeks, the grazing sheep had ingested too many radionuclides, and thus restrictions needed to be implemented to keep the contaminated meat from entering the human food chain. Later, it turned out that these restrictions would not last a mere couple of weeks, as initially predicted by the authorities, but some farmers would have to live with them for many years to come. However, even these clearly observable miscalculations by the state experts did not in any way lead to something like a British 'Chernobyl affair'. Such a thing simply never took place, and today only a few people even remember or know about the impact Chernobyl had on Britain. No state official was publicly accused of having purposely held back the truth about the fallout. Such an argument is almost completely absent in the British Chernobyl debate and this applies to institutions and individuals alike. There was no British Pierre Pellerin, which is to say, an individual accused of having communicated incorrect information. If somebody was indeed blamed for mismanagement of this emergency situation, it was the government and more specifically the MAFF. But the accusations were of incompetence, not of a deliberate cover-up. They gained in importance at local and regional levels, not at the national level. And after the government had delivered a detailed report to the House of Commons and created the UK Response Plan, all public debates on the issue were closed. Obviously, Britain was not without its critical voices, especially in May 1986: they blamed the British nuclear techno-political regime for its secrecy regarding the true impact of the accident, and for refusing to reconsider the new build plans in the face of the evidence that severe accidents were indeed possible. But these critical voices soon lost steam, or at least lost their public visibility. It seems the narrative released by the public authorities – 'what needed to be learnt had been learnt' – had supplanted these objections and views; now Britain was better prepared in the event such an accident should ever happen abroad again. Given that the integrity of the state experts had not been publicly questioned, but rather the government's management of the emergency, Chernobyl could be framed, from the state experts' perspective, as an experience from which to learn, which would in turn make it possible to improve the existing system.

In general, it seems like the British public placed more trust in their experts than in their politicians in May 1986. The public had in fact turned to these experts – in the form of 800,000 calls to the various agencies cited above – at a time when contradictory statements on Chernobyl had been released by members of national and local governments. This trust, at least with regard to radiation protection, does not seem to have been lastingly damaged, not even by the confusion about sheep farm restrictions. In France, however, the critical voices trusted neither the state experts nor the politicians insofar as they were considered to be one single entity within the French nuclear techno-political regime.

The one group of people who could potentially have turned the sheep restrictions into a British Chernobyl affair, the sheep farmers, did not however have their assumptions of bigger politics at stake in connection with Chernobyl. Rather, the sheep farmers in the Lake District related the restrictions to the Sellafield-Windscale complex. They considered the restrictions on sheep farms as a measure to cover up the regional impact of the Sellafield plant and the continuing impact of the 1957 Windscale Fire on the area. The Windscale Fire, not Chernobyl, was the discursive arena where nuclear state officials were publicly accused of having deliberately held back the truth about the fallout. Moreover, British nuclear critics considered the Chernobyl fallout as negligible compared to the level of threat represented by Sellafield.[289] Even after Chernobyl, Sellafield was still perceived as the primary incarnation of nuclear risk. For this reason, anti-nuclear plant protest in Britain has focussed on Sellafield and never on Chernobyl. The opposite occurred in France, where the *affaire Tchernobyl* became the focal point for anti-nuclear protest. In Britain, the Chernobyl experience was put in the context of the Sellafield experiences, whereas in France, it became the general frame used to criticize the *nucléocratie* as opposed to accidents in French nuclear plants or emissions from La Hague, the reprocessing plant in Normandy. In France, the Chernobyl experience also reinforced the profound mistrust many activists had of French nuclear experts, a mistrust that through the *affaire Tchernobyl* spread to a wide French public. Chernobyl had much less of a crucial impact on the credibility of public radiation safety expertise in Britain than it did in France. Even the open dispute between public experts and farmers in Britain did not lead to a lasting general distrust in nuclear experts. Wynne explains that the outcome of this dispute was found in the direct interaction of the two groups. This interaction implied that in spite of the confrontation, both sides were able to gain a better mutual understanding of the other's positions, working methods and shortcomings:

> This serendipitous and limited interaction improved the credibility of these scientists and of their associated institutions, even though such encounters revealed scientific uncertainty. The Institute for Terrestrial Ecology was most fortunate in this respect because, as a locally based institution, it had the closest such practical contact. Through the farmers' informal grapevine, it subsequently gained a reputation as being plainspeaking, open about uncertainty, independent and trustworthy.[290]

A similar dynamic enabled the rehabilitation of the MAFF in whose case 'the situation was rescued only by the mediation of local MAFF officials who were personally known and trusted by the farmers'.[291]

The fact that public experts had to publicly admit their ignorance of certain aspects of their field of expertise resulted in the restoration of their public credibility and authority.[292] Because events did not play out as the experts had predicted, they were perceived to be on a par with laypersons regarding their fallibility. In this light, it was even possible for the people most troubled by their miscalculations, the sheep farmers, to pardon them. The blame was articulated as 'they should have known better!' In France, however, the blame was expressed as 'they did know better!' The public experts were not only perceived to have failed in their task, but to have withheld their knowledge from the public, keeping it for a restricted circle of individuals. And though the public waited for an apology, this never came, since the state experts were convinced that they had not committed any errors in their risk evaluation. Therefore, it never came to a point, like in Britain, when the state authorities could enter into a dialogue at eye level with the concerned public. When this condition never materialized, the ground was effectively prepared for the *affaire Tchernobyl* to live on in the French public memory, and thus in the months following the Chernobyl crisis, the loss of trust was never resolved. When the French public experts believed the problem to be one of communication, this gap in trust, so to speak, widened even further. They believed, much like their British counterparts, that if they could only better educate the public on radiation risks, their credibility and authority would be restored. But the opposite proved to be true: the more they communicated, the more their communications were perceived and framed by their critics as a way to cover up their crime.

In France, the *affaire Tchernobyl* set the precedent for the institutionalization of nuclear counter expertise, namely the foundation of CRIIRAD and ACRO. In Britain, however, such strong and publicly visible counterexpertise on the impact of radiation did not exist in the civil nuclear enterprise. Nor were any trials conducted at the time to implement some such similar structure in relation to Chernobyl.[293] This implies that the personal histories of anti-nuclear power activists in Britain were not closely linked to Chernobyl, whereas in France, Michèle Rivasi, the founder of CRIIRAD, became one of the most well-known anti-nuclear activists and politicians in the Green Party. In addition, few journalists in Britain personally engaged in unveiling the 'truth' about Chernobyl's impact; from the early 1990s onwards, the topic of sheep farm restrictions almost completely disappeared from the newspapers, despite the ongoing enforcement. In France, on the other hand, several journalists felt personally betrayed by the information policies of the public authorities. They played a decisive role in creating the *affaire Tchernobyl* and kept its memory alive for years to come, especially on 26 April every year. This driving force to spur public commemoration of Chernobyl does not exist in Britain.

Regardless of all their differences in contextual setting, interestingly, the political parties in both countries did not play a prominent role in the early history of the Chernobyl debate, either as actors or as subjects of debate. This is not to say the political parties were entirely absent, in fact, the Green Parties in both countries took up the topic in order to position themselves against the government's nuclear policies. In the Labour Party, the topic of Chernobyl led to profound debates about the future role of nuclear power in British energy policies. However, in neither country did Chernobyl alter the general cross-party agreement on nuclear energy issues, which had existed since the launch of the national nuclear endeavour. This general cross-party agreement limited the opposition to this technology to the streets and was expressed by the public – thus the topic of anti-nuclear protest has always been closely associated with the rise of New Social Movements. According to many anti-nuclear campaigners, the established political parties adhered to the nuclear agenda of the *nucléocratie*, an agenda that was intrinsically opposed to anti-nuclear convictions. This perception was particularly strong in France and might explain why the *affaire Tchernobyl* revolved more around nuclear experts and less around the government. From an anti-nuclear activist's point of view, it was not surprising that the government acted in a way that protected the French nuclear enterprise. But within such a system, at least the public experts should be trustworthy. This would also explain why Pierre Pellerin soon became the main target of criticism. At the same time, this profound loss of trust in official nuclear experts paved the way for the new 'counter experts' to rapidly gain attention and credibility.

Based on these observations for France, where the crisis management was interpreted as a striking example of how the *nucléocratie* governed public policies, I would like to hypothesize why the British restrictions on sheep farms did not spark a similar British Chernobyl affair. Like in France, the topos of a 'nuclear technocracy' has been a central element of the British anti-nuclear discourse. It has been used not only for the military complex, but also in relation to the large-scale new build projects for nuclear power plants.[294] But, as Markku Lehtonen argued, 'in the UK, the power of the nuclear technocracy has declined since the split-up of the UKAEA [in 1971]'.[295] In addition, the successful campaign by British anti-nuclear power activists in the late 1970s against the government's plans to build a series of PWRs managed to weaken British nuclear technocracy not only institutionally, but also discursively. Thus, in the mid-1980s, this topos for the civil use of nuclear power had lost much of its *Wirkungsmacht* (power to influence). Not even the anti-nuclear power activists framed in this way the poor manner in which the sheep farm restrictions had been handled. The incidents involving the Sellafield-Windscale complex, however, were very clearly considered to have been wilfully covered-up by the British nuclear technocracy – at the very moment Chernobyl was drawing public attention to the issue of radioactive fallout in Britain. But somehow this was framed as a story from the past, and the narrative was not transferred or projected onto the situation caused by Chernobyl. In particular, the public experts' miscalculations regarding the scale and duration of the sheep farm restrictions were not interpreted in this way. The NRPB and the MAFF were indeed criticized for their incompetence, but their actions were not seen as an actuation of a bigger nuclear technocratic policy. In France, on the other hand, the statements released by the SCPRI and the Minister of Agriculture were considered to be part of a larger nucleocratic policy. As the example of the publication *Britain's Nuclear Nightmare* illustrates, there were still activists in the late 1980s who argued against the power of the British nuclear techno-political regime and blamed it for having a devastating impact on Britain's environmental and public health. But Chernobyl was barely part of this argument.

The scientists who evaluated the impact of the fallout in the hills were not considered part of the British nuclear techno-political regime. They were seen as independent experts who ensured that the radionuclides would not enter the human food chain. Although they had failed academically in their predictions, they were still considered to be independent from state nuclear policies. Thus, the French nuclear sector was perceived to be far more powerful and influential than the British nuclear sector and to have penetrated society far more intensively. From today's perspective, comparing the sheer size of the national nuclear fleets and the resulting shares of nuclear power in electricity generation – 80 per cent in France versus 16 per cent in Britain – it seems completely logical that the French nuclear sector is seen as more powerful and influential than the British. Again, from today's perspective, we might argue that there was simply no need to turn the miscalculations by British nuclear experts into a wider debate about power dynamics in the nuclear techno-political regime since the topic of nuclear power just did not play an important role in everyday life. However, this argument only partly conforms to the setting of 1986. Back then, France already had more nuclear power plants than Britain. But the predominant generation of electricity in France by nuclear power plants was quite a recent development that came about in the 1970s with the *Plan Messmer*. Like France, Britain had for some time considered itself a leading player in nuclear engineering. The nuclear power programme boosted national pride – the opening of the Calder Hall reactor in 1956 by Queen Elizabeth II is the most illustrative example. Furthermore, in 1986, Britain's plans to build – like France – an entire fleet of new PWRs, had been struck from the programme only a couple of years earlier. In the late 1970s, following the 1973–74 oil crisis, the CEGB and the British government had dedicated themselves entirely to nuclear power and considered it the future of electricity generation, the only way to 'keep the lights on'. The initial plans not to import this new fleet of reactors was yet another similarity with French nuclear policies. Britain had in fact invested intensively in developing its own reactor design. Although the plans for large-scale new build projects were eventually rescinded, the Thatcher government was very much pro-nuclear and pushed for the construction of new reactors on the existing sites at Sizewell and Hinkley Point. Plans for more new build were only abandoned in the early and mid-1990s, with the privatization of the British energy sector. Thus, in 1986, Britain and France were much closer in terms of their political elites' nuclear mind-set than today's shares of nuclear power in electricity generation might suggest. Therefore, it makes sense to ask firstly why the civil nuclear techno-political regimes were perceived so differently in France and Britain, and secondly, why the regime's power dynamics were associated with the official handling of the Chernobyl crisis in France, while this interpretative framework barely played any role in the public debate in Britain.

To explain this difference, we must look at the anti-nuclear protest's different focal points in each country. In France, the anti-nuclear power movement (also known as 'anti-reactor activism') was very strong, whereas the British anti-nuclear protest was far more concerned with nuclear weapons. And while there were also activists who directed their actions against nuclear power plants, they found themselves in quasi-competition with the CND for public attention, a competition the CND clearly won. Most British anti-nuclear campaigners had a more global vision as they turned their focus to the universal threat of a nuclear war. The issue of nuclear power plants was then grouped together with the threat of proliferation. Even though this issue was seen as a minor part of a larger enemy, the weapons-focussed British anti-nuclear protest did not entirely overlook the national civil nuclear sector but considered it in a more global context. This contextualization of the civil nuclear sector is represented in the book *Something in the Wind* by Mackay and Thompson: in which the authors used Chernobyl to underpin their position in the global nuclear disarmament discourse, not to frame their interpretation of the event as a direct argument against the British civil nuclear programme. Ultimately, what would be the benefit of phasing out nuclear power in Britain when the nuclear arms race continued to go unchecked at the global level? The national and global military nuclear techno-political regime was thus the focus of the British anti-nuclear campaigners' protest. This framing is present also in works by other critical authors: Hamman and Parrott, Haynes and Bojcun. In France, however, anti-nuclear protest did not primarily address the global scale. Therefore, Chernobyl was discursively embedded in the national nuclear power techno-political regime.

We can draw a broader conclusion that the dissimilar trajectories of the French and British Chernobyl debates were due to a different perception of the power of the countries' civil nuclear techno-political regimes. In France, Chernobyl's impact was considered at the national level within the discursive context of criticism of the *nucléocratie*. In Britain, no such interpretative framework existed for the civil nuclear programme, since anti-nuclear criticism had been predominantly directed against the military complex and focussed more on international relations than the national nuclear energy complex. Thus, the British Chernobyl debate put the accident in a global perspective, whereas the French Chernobyl debate focussed on the impact at home.

Notes

1. The RBMK is a graphite-moderated and light-water-cooled reactor developed in the Soviet Union and was only built at facilities in its (former) territory.
2. Most publications on Chernobyl give a detailed description of how the accident evolved. I therefore refrain from adding yet another description and provide just a brief account.
3. For more information on the resettlements and clean-up workers, see: Sahm and Sapper and Weichsel, 'Tschernobyl: Vermächtnis und Verpflichtung'.
4. The articles were accessed in the British Library's ProQuest Database, using the search key 'Chernobyl'. I chose the *Guardian* because it includes the widest range of opinions on nuclear power in its editorials. In the era researched in this book, it is also the British newspaper most dedicated to investigative journalism; its readership is mostly left-liberal oriented.

5. Fairhall, 'Radioactive Russian Dust Cloud Escapes'.
6. Ibid.
7. Ibid.
8. Walker, 'Silence Covers "Zone of Death"'.
9. McRae, 'The Soviet Nuclear Disaster'.
10. Walker, 'The Vision Consumed in the Fire'.
11. Tucker, 'Unquenchable Core of Fear'.
12. In 1957, this accident caused the release of substantial amounts of radionuclides into the environment. The areas most affected by the fallout were the immediate surroundings of the Windscale plant, part of today's Sellafield site located on the Irish Sea coast near the Lake District in north-west England.
13. Tucker, 'Unquenchable Core of Fear'.
14. Fairhall, 'Chernobyl Factors'.
15. *The Guardian*, 'A Disaster without Frontiers'.
16. Keel and Walker, 'British Evacuees Fly Home'; Walker and Valliamy, 'Airport Radiation Tests'.
17. White and Walker and Brummer, 'US Estimates up to 3,000 Victims'.
18. Hooper, 'Britain Sticks to Policy'; *The Guardian*, 'PM Rejects Sizewell Rethink Call'.
19. Brown, 'Labour at Odds Over Nuclear Power'.
20. Vletch, 'Parents Scour Chemists Shops'.
21. Fairhall and White and Tran, 'Radiation Cloud Reaches Britain'.
22. Ibid.
23. Brown, 'Labour at Odds Over Nuclear Power'.
24. Tucker, 'Europe Fears Effects of High Radiation'.
25. Stead, 'Radioactivity Alarms Scotch Makers'.
26. *The Guardian*, 'Baker Acts to Soothe Radioactivity Fears'.
27. Naughtie and Rose, 'MPs Told: There Is No Health Risk'.
28. Tucker, 'Levels Dropping Fast'.
29. Heath, 'Contaminated Rain Sets Welsh Alarms Ringing'.
30. Naughtie and Rose, 'MPs Told: There Is No Health Risk'.
31. Ibid.
32. Young, 'Crisis of Credibility'.
33. Interestingly, from the very beginning the issue of trust (in politicians and in public experts) and how to regain trust in their evaluations, became central to the Chernobyl debate. This is in line with Ute Frevert's observations. Frevert considers trust an 'obsession of modernity', a tendency in modern times to subsume progressively more relations under the discursive and emotional umbrella of trust and to consider trust in persons or entities as a major goal to achieve. Frevert, *Vertrauensfragen*.
34. Young, 'Crisis of Credibility'.
35. Ibid.
36. Taylor, 'Chernobyl Cloud Jams Whitehall Phones'.
37. This problem was addressed later through the UK Response Plan, which established rules for dealing with the effects of an overseas nuclear accident on the UK. The UK Response Plan framework, including the Radioactive Incident Monitoring Network (RIMNET) database, was implemented in the 1990s as the national radiation moni-

toring and emergency response system. I am grateful to Ian Fairlie for this information on RIMNET. For a proposed outline of the Response Plan, see: Department of the Environment, *Nuclear Accidents Overseas*.

38. Lashmer and McKie and Lean, 'Confusion in Britain'.
39. See: Stead, 'Scottish Fallout Four Times Critical Levels'.
40. Jennifer Brown states this number in 'The Impact of Television Coverage on the United Kingdom', 196.
41. The following section on France is a summary of the corresponding paragraphs in Kalmbach, *Tschernobyl und Frankreich*, 65ff. All translations from French sources by the author. The account is based on research in the Sciences Po Paris press clipping collection and TV news reports archived at INA. For a compilation of French newspaper reports on Chernobyl in 1986, see: Strazzulla and Zerbib, *Tchernobyl – Les médias et l'événement*.
42. France 3, *Soir 3*.
43. Tardieu, 'Catastrophe dans une centrale nucléaire soviétique'.
44. Croizé, 'Le syndrome ukrainien'.
45. Streiff, 'Le syndrome de Tchernobyl'.
46. *Le Quotidien de Paris*, 'Catastrophe nucléaire en URSS'.
47. Jamet, 'Le S.O.S de l'URSS'.
48. Ibid.
49. Clarke, 'Le Matin répond aux 10 questions'.
50. See: Vodnik, 'Le syndrome du nuage panique les Polonais'.
51. *Libération*, 'L'Europe dans le nuage'.
52. Here, 'radiation' always means 'ionizing radiation'. The terms 'radiation protection' and 'radiological protection' are used synonymously.
53. I only outline the institutions and responsibilities in the French radiation protection and safety sector, concentrating on SCPRI and IPSN in the French Chernobyl debate. For more detailed information on this sector between 1986 and 2006, see: Renaud and Champion and Brenot, *Les retombées radioactives*, 175ff.
54. For the history of this institution, see: SCPRI, *Organisation pratique de la radioprotection*.
55. When Pellerin died in March 2013, *Le Figaro* published an obituary entitled 'Tchernobyl: mort du Professeur Pierre Pellerin'.
56. This is a brief outline of how these institutions were restructured, not elaborating on institutions such as DSIN and DGSNR. For a comprehensive account of the creation of the IRSN and other institutions involved, see: Ministère de l'Economie, des Finances et de l'Industrie, *Une pièce essentielle dans l'organisation de la sûreté nucléaire*.
57. For a partial analysis of SCPRI's communiqués, see: Sezin Topçu, *L'agir contestataire*, 196ff. To determine who had access to which numbers at what time, requires a full comparison of the archives at SCPRI, newspapers and TV channels. This important aspect of the *affaire Tchernobyl* is a lacuna in French media history that merits further research.
58. TF1, Interview with Pierre Pellerin, in *13 h*, 29 April 1986, transcription of the original recording held by the INA, translation by the author (applies to all quotations from TV emissions in this book).
59. *Le Parisien*, 'Le nucléaire en France'.

60. A2, Interview with François Cogné, in *MIDI 2*, 29 April 1986.
61. Ibid.
62. The NRPB was instigated in 1970 and existed until 2005 when it became part of the newly founded Health Protection Agency (HPA). For a history of the NRPB, see: O'Riordan, *Radiation Protection*.
63. Fairhall, 'Radioactive Russian Dust Cloud Escapes'.
64. Fairhall and White and Tran, 'Radiation Cloud Reaches Britain'.
65. Travis, 'No Radiation Threat to the UK'.
66. Ibid.
67. *The Guardian*, 'PM Rejects Sizewell Rethink Call'.
68. Murtagh, 'Soviet Reactor "Safer than British AGR"'.
69. Ibid. The issue surrounding the definition of containment was not unique to the British debate and was also discussed at the international level, particularly within the IAEA.
70. Hooper, 'Britain Sticks to Policy'.
71. Clark, 'John Dunster CB ARCS BSc FSRP (1922–2006)'.
72. Wavell and Cook, 'Case of Missing Isotope'.
73. Naughtie and Rose, 'MPs Told: There Is No Health Risk'.
74. British Nuclear Energy Society, *Chernobyl: A Technical Appraisal*.
75. Ibid., 64.
76. Ibid., 67.
77. Ibid.
78. Ibid., 77.
79. Ibid., 86.
80. NRPB, *Preliminary Assessment of the Radiological Impact*. The authors are: M. Morrey, J. Brown, J.A. Williams, M.J. Crick, J.R. Simmonds, M.D. Hill.
81. Ibid., abstract.
82. Ibid., chapter 2.
83. Ibid., 23.
84. Ibid., 26.
85. Gittus et al., *The Chernobyl Accident and Its Consequences*.
86. A section of the chapter 'Nuclear Power Industry' discusses the CEGB and its role in the early Chernobyl debate.
87. Baker, 'Obituary: Lord Marshall of Goring'.
88. Ibid.
89. Gittus et al., *The Chernobyl Accident and Its Consequences*, 3.
90. In risk perception terms, this claim reflects the strategy of 'educating the public'. The public is thus considered to be contesting a technology due to a lack of or false information, and so providing more or better information will overcome their opposition. This presumes knowledge hierarchies between experts and the uninformed public and knowledge to be something produced in an expert circle, then passed to laypersons in a top-down-scheme.
91. Gittus et al., *The Chernobyl Accident and Its Consequences*, 7.11.
92. Ibid., 7.12.
93. These comparisons have been severely criticized: firstly, smoking is an active decision, whereas inhaling air containing radioactive iodine is not a choice (a letter to the editor

in the *Guardian* on 9 May 1986 sarcastically summarized this criticism: 'I have successfully given up smoking and am now trying desperately to give up nuclear power.'). Secondly, statistical comparisons are accused of creating unrealistic scenarios to distract from problematic individual cases. It is extremely unlikely, if not impossible, that a person would smoke 3/10,000 of a cigarette. The individual would probably smoke an entire cigarette on a daily basis, even more than one, and therefore run a higher risk of developing lung cancer. If this person was placed in a group of non-smokers, the average risk for each of this group would be lower than calculated for the smoker. This same logic sometimes calculated Chernobyl's long-term effects. The total radioactivity was divided among the total number of people affected by the fallout, namely everyone in Eastern Europe. Consequently, the calculated individual dose was very small. But, as with a smoker, this dose says little about individuals. For a general questioning of probabilistic risk assessment methods in the nuclear enterprise, see: Downer, *Nuclear Safety: A (Charlie) Brownian Notion.*

94. Using statistics in public discourse to illustrate 'facts' or 'truth' is a longstanding tradition. For the history of the political function of statistics in the Early Modern period, see: Behrisch, 'Political Economy and Statistics in the Late Ancien Régime.'
95. Ballard, *Nuclear Safety after Three Mile Island and Chernobyl.*
96. Ibid., v.
97. Ibid., 18.
98. Ibid., 9.
99. Ibid., 12.
100. Ibid., 22.
101. For an analysis of implementing this image, see: Mühlenhöver, *L'environnement en politique étrangère.*
102. Ballard, *Nuclear Safety after Three Mile Island and Chernobyl*, 23.
103. Ibid., 90–105.
104. Ibid., 354–89.
105. Ibid., 406.
106. DoE et al., *Levels of Radioactivity in the UK.*
107. In alphabetical order: Department of Agriculture for Northern Ireland, Department of the Environment, Department of the Environment of Northern Ireland, Department of Health and Social Security, Ministry of Agriculture, Fisheries and Food, National Radiological Protection Board, Scottish Office, Welsh Office.
108. DoE et al., *Levels of Radioactivity in the UK*, 7.
109. From a STS perspective, there are two different ways to interpret this form of communication with the public. Firstly, as a counterstatement to involving laypersons in radiation safety issues; the undecipherable numbers emphasized the technical knowledge required for decision-making process. Secondly, as a form of empowerment of laypersons; providing the data gave informed readers the opportunity to make their own calculations.
110. I was unfortunately unable to obtain this document apparently published by HMSO. However, Chris C. Park in his book *Chernobyl: The Long Shadow*, 104ff., includes interesting quotes: 'The report stated that there was no convincing evidence that public health had been put at risk, although it was likely that some lamb with radiation levels above the statutory limit had been eaten by the public after Chernobyl. Contaminated

lamb could have reached the shops by several routes, and the report concluded that "it must therefore be probable that some did so". However, Mr Jerry Wiggin, Conservative chairman of the committee, stressed that "someone would have had to have an exclusive diet of contaminated meat for a lengthy period for any harm to come about."' Park showed that despite severe criticism regarding the late revelation of a radioactive hotspot in north Yorkshire, 'the committee concluded that the Government has been on the whole successful in its objective of ensuring safe food in Britain, and that it had "got it right for nearly everybody"' (105).

111. House of Commons Agricultural Committee, *Chernobyl – The Government's Response*.
112. For the list of Post-Chernobyl research and development, see: Annex B to the memorandum.
113. Ibid., 37.
114. Apart from this comment by E.A. Ryder, I did not find other Nuclear Installations Inspectorate (NII) public statements on Chernobyl. The NII, once named by Lord Marshall of Goring the 'independent nuclear "watchdog"', was the UK's national nuclear safety inspectorate. In 2011, when the British radiation safety sector underwent a period of restructuring, the NII was incorporated in the newly founded Office for Nuclear Regulation.
115. Ryder, 'Regulation of Nuclear Power', 11.
116. Another option would be to draw a line between operators and inspectorates. But this line is problematic to define as nuclear institutions often controlled themselves, as exemplified by the IPSN, an inside-institution of the CEA.
117. Underpinning this issue of reactor design (a gas-graphite reactor developed in France versus a light-water reactor developed in the US) were two competing 'techno-political regimes', i.e. (very simplified) two different theoretical and practical approaches to the nuclear future of France. In her detailed analysis of the French post-war nuclear enterprise, Gabrielle Hecht identified a 'nationalist' regime in the CEA and a 'nationalized' regime in the EDF, see: Hecht, *The Radiance of France*. Markku Lehtonen, Claire Le Renard and Arthur Jobert identified a similar situation in the British case, where the CEGB favoured the American PWR design, while the AEA favoured the Advanced Gas-Cooled Reactors (AGR) developed in Britain. See for this interesting comparison: Lehtonen and Le Renard and Jobert, 'The Diverging Trajectories of Fast Breeder Reactor Development'.
118. EDF, *Rapport annuel 2008*.
119. A2, Interview with Pierre Tanguy, in *MIDI 2*, 30 April 1986.
120. EDF, *Tchernobyl: Dossier d'information*. The account of EDF publications is adapted from: Kalmbach, *Tschernobyl und Frankreich*, chapter 5.2.
121. EDF, *Tchernobyl: Dossier d'information*, 1.
122. Ibid.
123. Richard, 'Tchernobyl: Quelles retombées en France?'.
124. EDF, *Tchernobyl*.
125. Ibid., 1.
126. Ibid.
127. Ibid., 28.
128. Ibid., 34.
129. Ibid.

130. Ibid.
131. Ibid.
132. For the French case, see: Hecht, *Radiance of France*. In Britain, the Sellafield visitor centre was very popular. It was only closed in the early 2010s, due to emerging vigilance policies, fearing terrorist attacks.
133. Collier and Davies, *Chernobyl*.
134. Ibid., 1.
135. Ibid., 18.
136. Ibid., 20.
137. IBC Technical Service, *Nuclear Risks*.
138. Edmondson, 'The Chernobyl Accident and Its Implications for Operators', 1.
139. Ibid., 8.
140. Ibid.
141. Ibid.
142. Marshall, *The Chernobyl Accident*.
143. Ibid., 1.
144. Ibid., 3.
145. Ibid. (last sentence of the text).
146. All member institutions of the Watt Committee on Energy are listed on p. 141 of the quoted report and all its reports until 1988 on p. 142.
147. Worley and Lewins, *The Chernobyl Accident and Its Implications*.
148. The members of the working group (in alphabetical order): F.L. Allen, J.L. Bindon, R. Bulloch, A. Charlesby, D.R. Cope, P.M.S. Jones, J.D. Lewins, G. Lewis, J.G. Mordue, G.F. Oliver, G.K.C. Pardoe, P.D. Potter, F.B. Smith, G.N. Walton, N.G. Worley.
149. Worley and Lewins, *The Chernobyl Accident and Its Implications*, vi.
150. The other sections cover the design of the Chernobyl reactor (2); the evolution of the accident, based on the USSR report to the IAEA of August 1986 (3); UK and USSR reactor types (6); reactor operation and operator training in the UK (7); international dimensions of the implications of the Chernobyl accident for the UK (8).
151. Worley and Lewins, *The Chernobyl Accident and Its Implications*, 31ff.
152. Ibid., 32.
153. Ibid.
154. Ibid.
155. Ibid., 38.
156. Ibid.
157. Ibid., 39.
158. Ibid., 38.
159. Ibid., 95.
160. Ibid., 98.
161. Ibid., 61.
162. *The New York Times*, 'Chernobyl Trials Begin'. The accused were: Nikolai M. Fomin, Chernobyl's chief engineer; his assistant, Anatoly S. Dyatlov; Viktor P. Bryukhanov, plant director; Yuri A. Laushkin, senior engineer and inspector; Boris V. Rogozhin, shift director of Reactor 4; Aleksandr P. Kovalenko, head of Reactor 4.
163. British Nuclear Energy Society, *Chernobyl: A Technical Appraisal*.
164. Ibid., 1.

165. The dates are given in Nisbet and Woodman, *Options for the Management of Chernobyl-Restricted Areas.*
166. These numbers are taken from: Wynne, *Sheepfarming after Chernobyl*, 14.
167. Wynne, 'Misunderstood Misunderstanding'. This article was republished under the same title (an 's' was added to make 'Misunderstanding' plural) in 1996 in the book *Misunderstanding Science? The Public Reconstruction of Science and Technology* (edited by Alan Irwin and Brian Wynne). The 1996 version includes an additional section discussing the assumptions which shaped the official experts' scientific knowledge and the 'cultural issues' central to Wynne's analysis: trust and credibility. Wynne highlights the interplay between state experts and local farmers. To Wynne, the case of the Lake District farmers illustrates 'the unacknowledged reflexive capability of laypeople in articulating responses to scientific expertise' (301). Wynne's work has become a classic within STS. In their famous book *The Golem at Large: What You Should Know about Technology*, Harry Collins and Trevor Pinch chose Wynne's work on Chernobyl and Cumbrian sheep farmers to illustrate the mechanisms of knowledge production in science and technology. However, the sheep farm restrictions are barely known outside the STS-community and the success of Wynne's work has obscured the fact that other regions in Britain were also severely affected. This interest in Lakeland farms is certainly linked to the special position the Lake District holds among Brits as a nostalgic tourist and outdoor destination.
168. O'Connell, 'Letter to the Editor'.
169. For the way the media reports linked the sheep farm restrictions in the Lake District with Sellafield and the Windscale Fire, see: Cunningham, 'When Lamb Provides the Beef'; Lean, 'Nuclear Family Haunted'.
170. State experts dismissed this argument by referring to different isotope ratios of caesium 134 and 137 in the discharges from Sellafield and Chernobyl, see: Wynne, *Sheepfarming after Chernobyl.* The British Library catalogue lists a comparative study by the AEA on the Windscale and Chernobyl discharges. However, this item is unfortunately missing, and I was unable to consult it in another archive: Chamberlain, *Comparisons of the Emissions in the Windscale and Chernobyl Accidents.*
171. Wynne, *Misunderstood Misunderstanding*, 299.
172. For a map of the restricted areas in Cumbria and Wales, see: Nisbet and Woodman, *Options for the Management of Chernobyl Restricted Areas*, 2.
173. Heath, 'Farmers Take Chernobyl "Hostages"'.
174. Farmers' Union of Wales, *Chernobyl – the Government Reaction.*
175. Ibid., 6.
176. Ibid., 10.
177. Ibid., 13.
178. Ibid., 17.
179. Country Landowners Association North Wales Group, *North Wales Group Report.*
180. Ibid., 19.
181. Ibid., 9.
182. Ibid., 19.
183. Ibid., 2.
184. Ibid., 16: 'The problems of dealing with the long-term, contaminated peat areas and their side effects are outside the remit of MAFF. Advice to the Ministries concerned

can best be procured through a Scientific Inquiry by a qualified committee. The present policy of secrecy is unproductive. … Government must resist the temptation of banning sheep from grazing the peaty areas in order to overcome the problem. The removal of livestock from these areas will not only have adverse social and economic effects on the communities which derive their livelihoods from the mountain sheep, but also will result in the degradation of the landscape. It must not be forgotten that areas such as the Snowdonia National Park owe much of their beauty and grandeur to the management by farmers'.

185. Ibid., 20.
186. Ibid., 34.
187. See: Ardill, 'Chernobyl Leaves Shadow over 270,000 Sheep'; Lean and Heath and Siddell, 'Fell Lambs Born under N-Cloud'; Vulliamy and Heath, 'Chernobyl Controls on Sheep Extended'; Brown, 'Welsh Sheep Farmers Suffer'. According to the article 'Chernobyl Controls on Sheep Extended', the number of affected farms at the end of August 1987 was 345 in Wales, 69 in Scotland (of which 39 were not yet restricted in 1986), and 150 in Cumbria.
188. See: Jopling, 'Sheep Still Too Hot to Handle'; Halsall, 'New Controls Spur Fall-Out Inquiry Call'; MacDonald, 'High Radiation Levels Scar Rural Families'.
189. See: Johnson, 'Walker Pledges Review of Chernobyl Claims'.
190. Heath, 'Radiation Checks on People in North Wales'.
191. For an account of the British anti-nuclear power movement and protests at Torness in 1979, see: Welsh, *Mobilising Modernity*.
192. This different focus of the British and French anti-nuclear movements is analysed in: Chafer, 'Politics and the Perception of Risk'. Christopher Rootes provides a concise summary of British anti-nuclear power activism from its inception to the mid-1990s: Rootes, 'Britain', 21ff.
193. Friends of the Earth, 'The Accident They Said Would Never Happen'.
194. Friends of the Earth, 'A Positive Message'.
195. I am grateful to Walt Patterson for information on the anti-PWR campaign.
196. This issue is discussed in chapter 3 as it gained greater momentum within the climate change debate and led to the split in the British environmental movement over whether nuclear power is benign and 'green'.
197. The Greens, 'Windscale 1957'.
198. Greenpeace's 1987 report *Chernobyl UK* argued that a similar accident could happen in Hinckley Point B. I could not procure a copy, but Cutler and Edwards refer to it in their book *Britain's Nuclear Nightmare* (see later in this chapter).
199. I am grateful to Rebecca Johnson for sharing her thoughts on Chernobyl's impact on the CND campaign. There are various publications on CND, for its early history, see: Nehring, 'Cold War, Apocalypse and Peaceful Atoms'. For an account from within the CND, see: 'The History of CND' on CND's website.
200. CND, 'What If Nuclear War Was Just an Accident'.
201. CND, 'Sellafield, Libya, Chernobyl'.
202. I am grateful to Ian Fairlie for drawing my attention to this issue.
203. Turner, 'Let the Campaign Roll Again'.
204. Mackay and Thompson, *Something in the Wind.*
205. See for the END: Burke, 'END: Transnational Peace Campaigning in the 1980s'.

206. Mackay and Thompson, *Something in the Wind*, 1.
207. For further information on Zhores Medvedev, see chapter 2.
208. Mackay and Thompson, *Something in the Wind*, 7.
209. Thorpe, *Doc Chaos*.
210. Ibid., the book's back cover.
211. Ibid., 83.
212. For the history of the GSIEN, see: Topçu, 'Les physiciens dans le mouvement antinucléaire'. The GSIEN can be compared to the American Union of Concerned Scientists.
213. The French anti-nuclear movement, its activists as well as its transnational linkages, has been a prominent topic of research for sociologists, political scientists and historians. For classic works, see: Touraine et al., *La prophétie antinucléaire*; Nelkin and Pollak, *The Atom Besieged*. For more recent publications, see: Topçu, *La France nucléaire*; Tompkins, *Better Active than Radioactive*.
214. For the GSIEN's reactions to Chernobyl, see: Topçu, 'Confronting Nuclear Risks', 233.
215. Ibid., 235.
216. The following paragraphs are adapted from: Kalmbach, *Tschernobyl und Frankreich*, 72ff.
217. Tardieu, 'L'état de l'hexagone après le nuage'.
218. Hélène Crié, 'La France, lanterne rouge'.
219. Cited in: Tardieu, 'La France miraculée'.
220. TF1, Interview with Pierre Pellerin, in *13 h*, 10 May 1986. Presentation of the maps from minute 21 onwards.
221. For further reflections on the reasons for the French journalists' reactions, see: Kalmbach, *Tschernobyl und Frankreich*, 75.
222. A2, Interview with Alain Carignon, in *MIDI 2*, 11 May 1986.
223. *Libération*, cover page, 12 May 1986.
224. A2, 'Tchernobyl: rétro des mensonges'.
225. *Nucléocratie* is a specific French adaptation of the term 'technocracy'. It is used negatively to refer to the group of people holding key positions in the French nuclear sector, including the engineering and the administrative/policy side. The term was coined by French economist and journalist Philippe Simonnot in his book *Les nucléocrates*.
226. This dispute was noted and observed abroad, for example in a British newspaper: Page, '"Information withheld"'.
227. *Les Dernières Nouvelles d'Alsace*, 'Polémique sur l'information nucléaire en France'.
228. Pierre Guillaumat was an alumnus of the Ecole Polytechnique and an *ingénieur des mines*. He was also president of the CEA, later of the EDF and Minster of Defence, and thus considered the epitome of a *nucléocrate*. For his biography, see the website Les Annales des Mines 'Pierre Lucien Jean Guillaumat (1909–1991)'.
229. Chiquelin, 'Le clan de l'atome français'.
230. Morel, 'L'état français'.
231. Guihannec, 'Nucléaire: les silences de la France'.
232. These two books, one written from an anti-nuclear perspective – Yves Lenoir is a key figure in the French environmental and anti-nuclear movement; the other from a pro-nuclear perspective, are analysed and compared in: Kalmbach, *Tschernobyl und Frankreich*, 83ff.

233. Mould, *Chernobyl: The Real Story*.
234. This is a quote from the opening lines to Charles Dickens' novel *Hard Times*: 'Now, what I want is Facts. Teach these boys and girls nothing but Facts. Facts alone are wanted in life. Plant nothing else, and root out everything else. You can only form the mind of reasoning animals upon Facts: nothing else will ever be of any service to them'.
235. Mould, *Chernobyl: The Real Story*, ix.
236. Ibid., 256.
237. Ibid., 23.
238. Ibid., 35.
239. Ibid., 61.
240. Ibid., 48.
241. Ibid., 63.
242. Ibid., 63ff.
243. Ibid., 64.
244. Ibid.
245. Ibid., 72.
246. Ibid., 64.
247. Ibid., 77.
248. Ibid., 112.
249. Ibid., 105.
250. Ibid.
251. Ibid., 75.
252. Ibid., 128.
253. Ibid., 76.
254. Ibid., 192.
255. Ibid., 180.
256. Hamman and Parrott, *Mayday at Chernobyl*.
257. Ibid., 151.
258. Ibid., 170.
259. Ibid., 190.
260. Ibid., 212.
261. Ibid., 222ff.
262. Haynes and Bojcun, *The Chernobyl Disaster*.
263. Ibid., 204.
264. Ibid., 82.
265. Ibid.
266. Ibid., 204.
267. Ibid.
268. Ibid., 140ff.
269. Ibid., 149.
270. Ibid., 201.
271. Ibid.
272. Cutler and Edwards, *Britain's Nuclear Nightmare*.
273. Ibid., 164.

274. Margaret Gowing was the UKAEA's historian and archivist from 1959 to 1966 and wrote seminal works on British nuclear history in Second World War and the post-war period. For her life and work, see: Fox, 'Obituary: Professor Margaret Gowing'.
275. Cutler and Edwards, *Britain's Nuclear Nightmare*, 165.
276. Ibid.
277. Ibid., 166.
278. Ibid.
279. Ibid.
280. Ibid., 188.
281. Ibid.
282. Ibid., 187.
283. Ibid.
284. Ibid.
285. Ibid.
286. Ibid., 194.
287. A summary of this conclusion is part of: Bauer and Kalmbach and Kasperski, 'From Pripyat to Paris'.
288. Sezin Topçu stresses the 'singularity' of the French management of the Chernobyl crisis, see: Topçu, *L'agir contestataire*, 197. Although French authorities may have insisted longer than their British counterparts that there was no need for countermeasures, the initial rhetoric and reassurances were the same.
289. *Britain's Nuclear Nightmare* (discussed above) is a telling example of this view. It starts with a description of the landscape around Sellafield: 'On the western border of the Lake District a narrow single-track road winds through a series of hairpin bends over the crest of Corney Fell. In the shadow of the mountains, rolling pastures stretch along the shores of the Irish Sea. It is one of those special lakeland places the sight of which can give you a physical thrill. It can make you feel that life is worth living. Standing on the edge of the fells and surveying the coastline to the west, it is hard to believe that the beaches and estuaries are irredeemably polluted by one of the most poisonous of all man-made substances – plutonium. It is impossible to credit that the only places in the world more contaminated with radioactivity are the ghost town of Pripyat next to the burnt-out hulk of Chernobyl, and the remote sites around the globe where nuclear bombs have been deliberately exploded' (1). Although this region was one of the British areas most affected by the Chernobyl fallout, the authors only refer to the plutonium emitted by Sellafield, not the caesium and iodine deposited by rainfall in 1986.
290. Wynne, *Sheepfarming after Chernobyl*, 38.
291. Ibid.
292. This does not necessarily conflict with the British 'institutional mistrust' identified by Bickerstaff et al., 'Reframing Nuclear Power in the UK Energy Debate' and applied by Teräväinen and Lehtonen and Martiskainen, 'Climate Change, Energy Security, and Risk'. The notion of 'institutional mistrust' is linked to British nuclear policy debates for which Bickerstaff et al. show that the British public has more trust in markets than in their government and public authorities, a setting the authors call 'markets-know-best'. 'Institutional mistrust' might thus raise the question whether a privatized radia-

tion protection authority would have reacted better than a public entity. This does not oppose the thesis that the credibility of public radiation protection authorities could be restored through their interaction with the public.

293. The non-existence of Alternative Environmental Research Organizations (AEROs) in Britain might be due to a less centrally organized political system as well as strong stakeholder participation in decision-making processes through committees, inquiries and hearings. Especially the inquiries served as a 'security valve' (Walt Patterson) in the processes of identifying new zoning for nuclear plants (for a criticism of these forms of stakeholder participation, with a focus on the Sizewell B inquiry, see: Wynne, *Rationality and Ritual*). However, as AEROs are not yet a prominent research topic (except for West Germany: Rüdig, 'Mobilising Environmental Expertise'), any statement about the non-existence of AEROs in Britain is speculative.
294. Ian Welsh, in his book *Mobilising Modernity*, based on his 1988 dissertation 'British Nuclear Power: Protest and Legitimation 1945–1980', uses the term 'British nuclear technocracy'.
295. Lehtonen, 'Reactions to Fukushima in Finland, France and the UK'.

CHAPTER 2

1989–2005

Chernobyl Memory in the Making

This chapter traces the analytical structure of actor clusters and the key narrative elements used earlier. I focus primarily on the narrative element of personal affectedness, whilst the other two elements, radiophobia/apocalypse and anti-East European/anti-Soviet stereotypes, will take a secondary role. The reason for this is that perceived personal affectedness never dominated the British Chernobyl debate, whereas in France, it took centre stage, at the latest from 1996 onwards. The aim of this chapter is to closely examine this central difference between the trajectories of the British and French debates. When researching the narratives of Chernobyl debates in the UK and France, we should not limit our search to authors originating from these countries. From the early 1990s onwards – and in particular on the occasion of the fifth anniversary of Chernobyl in 1991 – external sources provided significant input to the British and French Chernobyl debates.[1]

The Transnational Chernobyl Debate

Zhores Medvedev

Zhores Medvedev was among the earliest critics from an East European background whose voice was heard in Britain. In 1987 already, he published an article in the *Environmentalist* criticizing the USSR's information policy regarding evacuees and people living in the most affected areas. This article was the basis for Mackay and Thompson's account of Chernobyl in their 1988 book *Something in the Wind*. In the English-speaking world, Medvedev had already established his credentials as investigative insider ten years earlier with his book on the Kyshtym accident;[2] thanks to his book, this accident at the military nuclear facility Mayak in the Urals became known to a broader public in the West.[3]

The fact that Medvedev had been exiled from the USSR in 1973 and had worked ever since as a research scientist for the National Institute for Medical Research in London, further strengthened his public credibility as an independent scientist. Thus, when he published his book *The Legacy of Chernobyl*[4] in 1990, it became one of the most important reference points in Britain. In it, Medvedev describes in detail the potential impacts of the Chernobyl fallout on people, plants and animals. To write this book, Medvedev had not undertaken research on Chernobyl himself, but had carefully studied the various papers and reports published on the topic both in the East and in the West. The result was a concise account of the state of the art in Chernobyl research, including his harsh criticism, for instance, of the unnecessarily high levels of radioactivity to which the emergency workers were exposed. Regarding the death toll, Medvedev refrained from taking a clear stance on numbers because: 'There is insufficient relevant information to enable an accurate assessment of the health consequences'.[5] Instead, he discussed the various reports published thus far. For the highest estimate, he presented a study by R.E. Webb published in the *Ecologist* in 1986, which predicted 280,000 extra cancer deaths worldwide.[6] However, Medvedev did criticize the fact that the author had 'consider[ed] radiation-related cancer deaths' only. From Medvedev's perspective – which was the point of view of a specialist in ageing research – 'the general reduction in life expectancy as a result of radiation exposure'[7] is what should have been researched. Furthermore, Medvedev pointed out 'odd' discrepancies in the debate on Chernobyl health effects, for instance: 'it should be stressed that the initial Soviet and IAEA reports on the level of radiation exposure of the population and the projected health risks made in 1986 were much higher than the figures given later. The origin of the confusion has never been explained'.[8] And last but not least, Medvedev linked his discussion of the various Chernobyl reports and accounts with statements on the wider energy policies discourse: 'The heated debate about possible future health effects of Chernobyl is understandable. … If the lowest estimates are right, the safety record of nuclear energy remains better than that of coal, oil and hydroelectric power. But if the highest estimates prove accurate, the outlook is vastly different'.[9]

Medvedev's narrative was mainly a critical account of the official evaluations. But his criticism also looked beyond Chernobyl, which he did not see as a single, isolated event. He considered the accident as belonging within the larger framework of Soviet energy policies and questioned many of the facts the international nuclear community had thus far agreed on as part of its evaluation. Medvedev did not present a concrete counter narrative to this official evaluation. His aim was rather to identify those areas where scientific knowledge was still lacking and highlight the issues that arise when existing knowledge is biased. In doing so, *The Legacy of Chernobyl* opened up the Chernobyl debate to other voices: a respected scientist with no connection to the nuclear sector making clear, comprehensible statements about problems with the official reports; it was a clear invitation to other 'counter experts' to join the debate.

As important as Zhores Medvedev's book was for the British Chernobyl debate, very surprisingly, it was never translated into French. Not even the English version is available in the French National Library. This is even more surprising given that Medvedev's book on Kyshtym was translated in 1988; Roger and Bella Belbéoch were the translators. Thus, *The Legacy of Chernobyl* is a particularly striking example of the extent to which national Chernobyl debates in Western Europe were separated by linguistic borders.[10]

Grigori Medvedev

Another Medvedev deeply influenced the French Chernobyl debate: Grigori Medvedev. His book, *The Truth about Chernobyl* was first translated from the Russian original into French in 1990,[11] before the English edition was published in 1991.[12] Grigori Medvedev was a high-ranking Soviet nuclear engineer, who had worked at the Chernobyl construction site in the 1970s and was involved in the 1986 investigation into the causes of the accident. In his book, Medvedev presented a detailed account of how the accident unfolded; his narrative would become the main reference for descriptions of the accident in years to come. But Medvedev did not stop at a detailed account of the chain of events leading up to and including the accident; he enriched his story with sharp criticism of the Soviet nuclear programme. According to him, the staff at the plant were clearly responsible for the events of 25 to 26 April: 'Toptunov and Akimov, who came on duty that night, as well as the operators and all the preceding shifts on 25 April 1986, failed to show the proper sense of responsibility and blithely proceeded to commit serious breaches of the nuclear safety regulations'.[13] However, the blame could not be placed on the staff alone; the deeper problem lay in the general logic of the nuclear system's functioning. It is worthwhile quoting the opening lines of Medvedev's final chapter 'A New Culture for the Nuclear Age' addressing his wider agenda:

> But so much still remains to be done! What further lessons still need to be learned! What battle must be fought in order to make our earth truly clean and safe for life and happiness! Meanwhile, the nuclear bureaucrats are not asleep. Though somewhat bruised by the Chernobyl explosion, they are once again rearing their heads, praising the completely 'safe' power of the peaceful atom, while not forgetting to cover up the truth. For it is not possible to sing the praises of the peaceful atom unless the truth is covered up.[14]

Medvedev's account describes the chain of events but not the impact of Chernobyl's fallout. *The Truth about Chernobyl* was not a counter narrative to any official report on health effects, but to the early official reports, also in Britain, that described the emergency actions as well planned and coordinated. Unlike the official reports filled with scientific data, Medvedev provided a personal story of the event: his was a minute-by-minute or even second-by-second account of the plant workers' and firefighters' struggle for survival, largely written like a play with dialogues. Transcriptions of eyewitness accounts complement the author's descriptions.

The material Medvedev collected for *The Truth about Chernobyl* and the reactions following its publication – including the attacks directed against him – were enough for a second book: *No Breathing Room: The Aftermath of Chernobyl* was published just two years later.[15] In *No Breathing Room*, Medvedev revealed the story behind his first book: how he had become an anti-nuclearist, his writing process, the difficult search for an editor and so forth. Medvedev underlined that the criticism he expressed in *The Truth about Chernobyl* had not been directed against the Soviet nuclear system alone. In this regard, the fall of the Soviet Union was for him a pivotal moment: 'Perhaps the most telling question of all: is the Western capitalist society really a suitable model for the former Soviet regime to follow – that is, can it be stated honestly that many of the problems that led to the Chernobyl catastrophe are not also present in the West?'[16] Medvedev's reasoning openly contradicted the West European public authorities' narrative that such an accident cannot happen here.

The introduction to Grigori Medvedev's second English Chernobyl book was written by David R. Marples. Professor at the University of Alberta and specialist in Ukrainian, Belarusian and Russian history, Marples was one of the first scientists in the humanities to conduct research on Chernobyl.[17]

Alla Yaroshinskaya

Another author who attempted to lift the veil on Chernobyl was Alla Yaroshinskaya in *The Forbidden Truth*.[18] Interestingly, the 1995 English edition of her book is not a translation from the Russian original, but from the French version, released in 1993.[19] Journalist Yaroshinskaya describes how information on Chernobyl had been systematically held back by the Soviet state. The following quote summarizes her argument: 'The most dangerous isotope to escape from the mouth of the reactor did not appear on the periodic table. It was not "Cs-137". It was "Lie-86". A lie as global as the disaster itself'.[20] In the US edition, published a year later by the University of Nebraska Press, Marples again provided an introduction. A foreword was added by John Gofman – former professor at UC Berkeley and one of the most prominent voices in the debate on the health effects of low-level radiation exposure. In 1992, together with Yaroshinskaya, Gofman was awarded the Right Livelihood Award – also referred to as the Alternative Nobel Prize – for his work on the health effects of ionizing radiation. Confirming claims in his own research, Gofman's introduction to Yaroshinskaya's text called for independent research into the health effects of radiation. He considered the current research policies – governed by 'a market eager for medical-unknowledge in the field of health consequences from nuclear (and other) pollutants',[21] – typically had the same dynamics Yaroshinskaya had uncovered with Chernobyl: 'In lavish sponsorship of scientists, engineers and physicians in innumerable institutions worldwide (including many halls of academe), the governments have a wish-list for the outcome of radiation research (into Chernobyl for instance). … Best of all would be the finding that a little extra radiation improves human health'.[22]

Vladimir Chernousenko

A similar argument about the true health effects of Chernobyl being systematically covered up, was put forward by Vladimir Chernousenko. Chernousenko had been to Chernobyl several times after 1986 in connection with his scientific work for the Ukrainian Academy of Sciences. His book *Chernobyl: Insight from the Inside*, published in 1991, aimed to debunk twenty-one 'myths of Chernobyl'.[23] These myths included statements like: 'The design of the RBMK-1000 reactor is impeccable. It was the operating staff that caused the explosion'; 'Only 31 people died as a result of the accident and the clean-up operations'; or 'The doses which people have received while living in contaminated areas will not have genetic effects'.[24] All in all, the list of the myths could be roughly equated with the early Chernobyl evaluations by pro-nuclearists. Chernousenko aimed to deconstruct these myths using a broad selection of mostly unpublished material, and interviews with eyewitnesses or people living in the most affected areas. He focussed on the health effects in these areas and described them as devastating: 'There are practically no healthy children in any of the towns or settlements there. The appearance of ambulances on school grounds is a common sight. Children faint from weakness right in the classroom'.[25] To underpin this argument, he quoted the Union of Chernobyl Liquidators: there were between 650,000 and 1,000,000 'liquidators' whose radiation-related illnesses were being denied and negated. According to Chernousenko, people had very good reason to be afraid: 'Is it any wonder that these "radiophobic" people think a crime was committed by those who made the negligent decision to suppress important, truthful information'.[26] His account is particularly interesting because he did not limit the Chernobyl health effects to cancer. In the chapter 'Doctor, Will I Live?', he presented an entire list of illnesses observed in children, including diseases of the liver, nose and throat as well as birth defects. The adults' state of health was no better: they suffered from gastritis, colitis, swellings, cardiovascular problems and other ailments.[27] Chernousenko did not claim these various illnesses were induced directly by radiation. But from his narrative, it would take only one, minute step to make this link; as did Belarusian doctors Nesterenko and Bandazhevsky, whose work I discuss in chapter 3.

Svetlana Alexievich

Svetlana Alexievich joined the transnational Chernobyl debate a few years later than the other authors. Her famous book *Voices from Chernobyl* was originally published in Russian in 1997 as *Chernobyl'skaia molitva*. In it, she gives a voice to the people evacuated in 1986, family of sick patients and deceased 'liquidators' and people who returned to their villages or moved to these regions because they considered them safer than their homelands. The book's worldwide success is mainly thanks to the artistic quality of the narrative, inspiring many types of artists: playwrights and theatre groups in particular for stage adaptation. However, *Voices from Chernobyl* is generally seen as documenting the situation around Chernobyl rather than a literary work. Consequently, quotations frequently feature as captions or explanatory texts in other books and exhibitions. Similarly, the makers of the successful HBO miniseries *Chernobyl* drew heavily from the book to create the series' story line. *Voices from Chernobyl* is a prime example of portraying Chernobyl as an apocalypse – the French translation even incorporates apocalypse in its title. The book's different reception in France and Britain is striking: the extremely positive response in France might be because of many people's perception of sharing a common destiny with those in the most affected areas of Eastern Europe. In Britain, Alexievich's book was published in 1999,[28] but unlike in France, it was neither promoted by book clubs, nor reprinted, and never became an important influence.[29]

Media Reporting on the Accident's Tenth Anniversary

In 1996, the British media did not allocate much space to the accident's impact. According to Lynn Frewer et al. in their study on Chernobyl media reporting in different European countries, the absence of this aspect in Britain was mostly due to the intensive reporting on the 'Dunblane tragedy'[30] and the BSE (commonly known as mad cow disease) crisis: 'Both events dominated media coverage during the time period covering the Chernobyl anniversary in the UK, and it is likely that the total amount of reporting about Chernobyl was reduced'.[31] However, the authors did not consider competing news events to be the predominant variable that dictated Chernobyl coverage in the national press. The authors of the British study stated that 'this reporting appears to reflect the degree to which a country is affected by the hazard (in terms of risk exposure and associated effects, such as economic impact)'.[32] This statement implies that neither Frewer's research group nor British journalists considered the radioactive fallout a major hazard to Britain, despite its having led to the massive restrictions on sheep farms, or at least not a major hazard compared to others such as BSE.

The situation in France at the time was quite different. Although Chernobyl was not a major news item in 1996, newspapers reported on the accident's impact. The articles also referred to the *affaire Tchernobyl*, reminding readers of the government's 'cover-up' of the fallout in France. The issue of its health effects gained extensive coverage in December 1996 when a wild boar in the Vosges Mountains in eastern France was found to have significant levels of contamination. In addition, after detecting an increase in a rare thyroid disease in children on the island of Corsica, local doctors, suspecting a link with radioactive iodine from the fallout, called for an epidemiological investigation. Corsica became a focal point in the French Chernobyl debate, particularly because the first maps delineating radioactive contamination in France did not include the island – which led to it being named the 'forgotten island'.

While speculations about Chernobyl health effects were already circulating in France in 1996, this was basically a non-issue in Britain. This makes the contents of a short article in the Scottish tabloid newspaper the *Daily Record* on 23 April 1996 all the more interesting:

> Experts have ruled out any link between cancer in the Western Isles and the Chernobyl disaster. A Benbecula doctor claimed a big rise in cases could have been caused by radiation fall-out from the atomic plant in the Ukraine. But a study by Western Isles Health Board said there was no link – and revealed that cancer rates in the islands were below the Scottish average.[33]

The case of Benbecula, a Scottish island situated 20 km west of the Isle of Skye, deserves attention as it demonstrates how the issue of Chernobyl-induced health effects was navigated in Britain. Barely a month earlier, on 31 March, the *Independent* informed readers that

> a sharp rise in cancer cases on a Scottish island is being linked to radioactive fallout from the Chernobyl explosion a decade ago. Doctors on Benbecula in the Outer Hebrides say the number of cancers has more than tripled in the past 18 months. They are demanding urgent investigation of the alarming rise, which they believe could be the result of people eating contaminated home-grown vegetables, and locally produced mutton, venison, and seafood, over 10 years.[34]

The fact that parts of England, Northern Ireland, Scotland and Wales had received significant amounts of radioactive fallout in May 1986 was not novel in 1996. The extensive restrictions on sheep farms (many still in place in 1996) had been a visible and direct result of this fallout. But what was new in the case of Benbecula was that local doctors linked this fallout to an increased incidence of cancer[35] – an increase they faced in their daily work and for which they had no obvious explanation. The doctors asked themselves: how did the island differ from other places regarding environmental factors, and what about the environment had changed that could trigger such a change in recorded cancer rates? Making the connections, the local doctors remembered the news reports of 1986 about the radioactive fallout, which had declared this fallout to be particularly high in the Western Isles due to heavy rainfall. They then wondered whether the consumption of contaminated home-grown produce and the inhalation of radioactive particles could have caused the local increase in cancer rates.

There are some remarkable similarities between the cases of Benbecula and Corsica. The debate on the health effects from ingesting contaminated home-grown foodstuffs in Corsica was also started by local doctors, having observed higher cancer rates in their patients. But in sharp contrast to the 'forgotten island' of Corsica, Benbecula did not instigate a nation-wide search for Chernobyl victims. The Benbecula story appeared for a few more days in British newspapers – the *Independent*,[36] the *Daily Record*,[37] the *Times*[38] and the *Herald*[39] – then disappeared forever. The only exception was an article on 23 April 1996, around the anniversary of the Chernobyl accident, in the *Daily Record* (cited above), saying that any link between the increase in cancer on Benbecula and Chernobyl had been ruled out.

It is not my intention to delve deeper into the medical details of these two cases, other than to point out an important difference between the two: in Corsica the debate was about thyroid cancer in children, whereas the doctors on Benbecula observed lung and bowel cancer in men in their 40s and 50s. The International Physicians for the Prevention of Nuclear War – one of the most critical 'counter expert' groups on the health effects of low-level radiation – would agree that thyroid cancer in children is far more likely to be linked to ingesting radionuclides than lung and bowel cancers in grown men. As average newspaper readers would probably not know these differences, we could expect that French reactions to a headline such as 'Chernobyl Causes Cancers in Corsica!' would be comparable to British reactions to a headline proclaiming 'Chernobyl Causes Cancers on Benbecula!'. Yet, the reactions in the two countries were poles apart. In France, Corsica became the linchpin of the debate on Chernobyl personal affectedness, while in Britain, the topic of personal affectedness in the Western Isles disappeared as rapidly as it had appeared, fading into history when reporting on Benbecula ceased.

This discrepancy comes down to the role played by state experts in the two countries, or more specifically, the confidence people placed in them at the time. In Britain, trust in official radiation protection experts was not profoundly questioned or challenged during the Chernobyl crisis, whereas the French public's trust in its institutions as well as individual staff members, was badly damaged. Therefore, in France, every study undertaken, and every report published by radiation and health protection agencies on the topic of Corsica only added fuel to the fire of the Chernobyl debate. Critics perceived any release of official evaluations as an attempt on the part of the public authorities to disguise and bury the true health effects in statistics; such calculations would not consider individual cases and therefore negate the existence of specific risks associated with certain diets. In Britain, however, the opposite happened: the Benbecula debate ended the minute that Dr Robert Kendell, Chief Medical Officer of Scotland, released his statement that it was 'exceedingly unlikely' Chernobyl was responsible for the increase in cancer.[40] No calls for proof of the statement were made, nor requests to await the final expert report, published three weeks after Kendell's declaration.[41] By that point, the report's news value was so low, it was not even mentioned in the *Daily Record* article on 23 April 1996. Although sceptical of the health officials dismissing the case, even the doctors who had started the debate did not defend their theory of a link between Chernobyl and the increase in cancer.[42] On 4 April 1996, the Glasgow-based newspaper *Herald Scotland* published an interview: Dr Francis Tierney

> said that he and his partner, Dr Andrew Senior, had not sought to make any connection with Chernobyl. He blamed the media for exaggerating the connection because of the forthcoming 10th anniversary of the explosion. Dr Tierney said he regretted the board's haste in dismissing radiation as a possible cause of any rise in cancer cases. He questioned why the board insisted that radiation caused only thyroid and leukaemia-type cancers, adding: 'What evidence is there for that? I certainly do not accept that theory.'[43]

But rather than looking further into Dr Tierney's doubts about the Health Board evaluation – which would have involved further reporting on intensive debates in the medical sciences about low-level radiation – the file on Chernobyl's detectable cancer effects in Britain was so firmly closed, that it did not even reopen on the tenth anniversary of Chernobyl a few weeks later.

Public Authorities

In addition to improving the man-machine interface, another British sphere that responded actively to Chernobyl was research into the transport and deposition of radionuclides in the air, water and soil. Several state agencies, universities and private research institutes were involved. The projects were supervised by the MAFF and often the wider public had access to the study results through government publications.

Not one of these studies held by the British Library commented on potential health effects of the Chernobyl fallout in Britain. The authors stuck to their research topic and all their statements adhered strictly to the precisely defined parameters of their study objectives. This is a markedly different approach to what we find in the French reports. French state experts nearly always included a section on the debate regarding French Chernobyl victims. This issue was simply not raised in British publications, a silence that is all the more surprising given that a probabilistic increase in cancer deaths due to the fallout was already being discussed by the spring of 1986. However, the estimate released by the NRPB in 1986, that the fallout would result in a few tens of deaths in Britain, was neither challenged nor taken up in the years to come. Not even the report titled *The Transport and Deposition of Airborne Debris from the Chernobyl Nuclear Power Plant Accident with Special Emphasis on the Consequences to the United Kingdom*[44] – which contained the very word 'consequences' in its title – made the slightest mention of how these debris could affect public health. This report provided in-depth information on the quantity of radionuclides deposited in Britain, as well as the where and when; it included detailed maps and described the progression of the plume over the UK in increments of several hours.[45] The report went on to present model calculations, combining data from meteorological observations of rainfall and air pressure with measurements of deposition on grass that government departments and the NRPB had collected.[46] From these calculations, the researchers generated the total deposition on the UK.[47] The authors then addressed the outcome of this deposition, stating in the chapter on 'Levels Observed in Cow's Milk', that 'the deposition of I-131 and Cs-137 led directly to incorporation in the human food-chain'.[48] But the report stopped there. It did not discuss the behaviour of the radionuclides once they entered the human food-chain; it did not mention the potential total amounts of radionuclides individuals may have ingested, nor contemplate the potential health effects of this food contamination. Similarly, the section on 'Longer-Term Agricultural Effects' stated that, regarding sheep farms, 'even 2 years after the event, levels remain sufficiently high for continued restrictions in some areas'.[49] Again, the authors did not elaborate on any risks linked to the contamination of these animals. Their only focus was the composition of the soil in these areas: poor acidic soil that did 'not lock-in the free caesium so that it becomes unavailable to the vegetation'.[50] The report identified that the key issue was the unforeseen behaviour of the soil regarding radionuclide capture, not the radioactive contamination itself. The authors did not judge the radioactive fallout from Chernobyl to be a major hazard, even not for the most affected regions in Eastern Europe. This can be deduced from their report's first chapter, 'The Accident at Chernobyl', in which they stated well-known numbers: 'Two people died immediately and 29 others died shortly afterwards as a result of severe radiation injuries incurred trying to contain the accident, whilst some 200 other people, mainly station personnel, sustained serious injuries resulting from being exposed to very high levels of radiation'.[51] There was no mention of long-term health effects. They concluded that the Chernobyl fallout was first and foremost an interesting subject and means of improving existing models: 'The experience gained from the behaviour of the radioactivity released from Chernobyl into the atmosphere and subsequently deposited on the surface is proving invaluable in the preparation of models, monitoring networks and other procedures for use should another major accident ever occur in the future'.[52]

These improvements in existing models would prove invaluable not only in the event of a future nuclear accident as far away as the Ukraine: the knowledge gained from Chernobyl would be even more crucial for the survival of British agriculture, should a severe accident occur in Western Europe, even on British soil. Thus, the public authorities' objective with personal affectedness research was less about evaluating the impact of Chernobyl on Britain and potential health effects, and more about using the accident as an opportunity to conduct experimental research on the effects of nuclear accidents, more precisely, the behaviour of nuclear fallout in general. This was evident in a joint AEA and MAFF publication. The report *Hot Particles from Chernobyl* stated 'the release of hot particles in a severe nuclear accident is shown to be an important factor to be taken into account in nuclear accident contingency planning. Their long-term behaviour in soils, especially with reference to migration and soil-to-plant transfer may be dominant factors in the feasibility of land reclamation'.[53]

It is interesting to compare this British perspective on the benefit of Chernobyl research to the French situation. In France, too, Chernobyl was considered an interesting source of data, which could refine models based on data for instance collected after Hiroshima and Nagasaki – an approach common to the international nuclear enterprise. But the experts' approach to Chernobyl as a source of learning evoked a different public reaction in France and Britain. While the British did not raise any major critique, French anti-nuclear activists were harsh in their criticism. French discussions over the CORE programme are a telling example. This programme had been initiated by French institutions to monitor and improve the living conditions of people in the most contaminated areas of Eastern Europe. The critics felt that the real motivation behind this programme was to create a training ground for French radiation protection experts to test the practicability of their guidelines in real life conditions.[54] They also accused the programme of being an attempt to prove that living in areas with high levels of radioactive contamination was indeed possible, and thus severe nuclear accidents would not be as fatal as most people believed. In Britain, however, where national radiation protection authorities openly communicated that their main interest in Chernobyl research lay in preparing themselves for future accidents, this stance never led to wider criticism.

The wide range of studies undertaken by various British research institutes conformed to this approach of using Chernobyl to review existing models. These studies investigated the transport and deposition of airborne radionuclides, transport in water systems and uptake mechanisms in plants and animals, as well as transport mechanisms in different types of soil. The systematic protraction of the sheep farm restrictions was proof that the existing models were inaccurate. Thus, sizeable amounts of money were spent on measuring and collecting samples then adjusting databases and algorithms. To give readers an idea of the scope of this research, I mention some of these studies. They are mostly pages and pages of measurements followed by a one or two-page conclusion in which the authors do not refer to broader issues such as potential health effects of the researched radionuclides. One 1990 report by the Department of Agriculture and Fisheries for Scotland *Chernobyl Accident: Monitoring for Radioactivity in Scotland*[55] analysed how much radioactivity may have been taken up according to the amount of beef or sheep meat consumed. However, it said nothing about possible (or probabilistic) health effects of this uptake. Other reports include: the MAFF, *Radionuclide Levels in Food, Animals and Agricultural Products 1987: Post Chernobyl Monitoring in England and Wales* (1988);[56] Her Majesty's Inspectorate of Pollution, *Chernobyl: Evaluation and Radiological Assessment of Factors Affecting Drinking Water Supplies* (1989);[57] the AEA, *A Survey of Radioactive Caesium in British Soils* (1990);[58] Her Majesty's Inspectorate of Pollution, *Measurements of Radioactivity from Chernobyl in Population Groups in Scotland* (1991);[59] the MAFF, *Radioactivity in Freshwater Systems in Cumbria (UK) Following the Chernobyl Accident* (1989);[60] the AEA/MAFF, *Hot Particles from Chernobyl: a Review* (1991);[61] and the AEA, *Transport Mechanism and Rates for Long Lived Chernobyl Deposits in mid-Wales* (1991).[62]

Some might ask: why would scientists have commented on possible health effects of the radiation in these reports? Given that the studies were rooted in the natural sciences and ideated to examine one specific aspect of a larger issue, if the authors had commented on aspects other than their specific research topic, their statements would have amounted to ungrounded speculation. The point here is not to criticize the studies or the authors, but to highlight that their narrow focus reflects the perception of the fallout as a neutral object of scientific interest. This is remarkably different from the French case, where research on the Chernobyl fallout was anything but a neutral topic from the outset.[63] As a result of the 1986 *affaire Tchernobyl*, any French research on Chernobyl had to be on one or the other side of the debate: either with the official experts, defending their expertise against the 'irrational behaviour of laypersons', or with the 'counter experts', whose research aimed to uncover the '*mensonge d'état*', and was frequently linked to an overarching criticism of the French nucleocratic system. The amount of literature on Chernobyl increased from the mid-1990s, at a time when the debate on French personal affectedness really gained momentum. In Britain, however, where this debate was marginal, research activities came to a halt in the early 1990s. Once everything had been said about the transport and deposition of radionuclides, the British state agencies did not release further publications on the impact of Chernobyl. What this means is that research into personal affectedness in Britain had come to an end before it had even begun in France.

The fact that health-related personal affectedness never became prominent in the UK may also be linked to British doctors' perception of the fallout's health impact. The cases of Benbecula and Corsica underline the importance of medical practitioners in this debate. In both instances, local doctors highlighted a potential connection between a rise in cancer rates and the Chernobyl fallout. In France, Corsican doctors and their evaluation of the health impact took centre stage in the Chernobyl debate, whereas this topic was not further investigated or corroborated among medical practitioners in Britain. One interesting source that supports this observation is an Institute of Physical Science in Medicine 1986 publication with the proceedings of two round tables that brought together hospital physicians from all over the UK to discuss their experience with the anxiety that people expressed in early May 1986.[64] It is worth noting that none of the papers at these round table meetings dealt with possible health effects of the radioactive fallout. They instead focussed on strategies to reassure anxious people that their health was not threatened. A paper by P.P. Dendy et al. on 'Whole Body Monitoring in Cambridge' illustrates how problems were prioritized: 'Most people monitored were reassured and were most grateful for our help. However, we were inundated by inquiries, which demonstrated that the amount of reassurance required by the public is beyond the resources of an average Medical Physics Department to provide'.[65] Thus, it was people's fear of the fallout, not the fallout itself that British doctors considered problematic.

It is important to point out that although British nuclear and health experts primarily considered Chernobyl fallout an interesting research topic or a public anxiety problem, public authorities were confronted with a decisively concrete challenge: radioactivity in sheep. Heavy rains in the west and north-west of the country in early May 1986 caused high levels of radionuclides to fall on the ground, and in the highland regions, the radionuclides were taken up by grazing sheep, leading to sheep farm restrictions in June 1986. Contrary to the forecasts by British radiation protection agencies, these restrictions would not be lifted after just a couple of weeks. In 1999, 13,000 sheep in Cumbria and 180,000 sheep in North Wales were still subject to the restrictions because their radioactivity levels were too high for them to go to market.[66] These restrictions were continually extended in time due to the sheep's continuing uptake of radionuclides. The official assumption was that the radionuclides would be rapidly locked into the soil and no longer taken up by the vegetation growing in the topsoil, and so would not be taken up by grazing sheep. However, the radioactive particles in the highland soil behaved quite differently than predicted. The experts' surprise at this phenomenon is reflected in the number of research projects conducted to study the transport and deposition of radionuclides in soils.

With the exception of British natural science academia and the farming community, the topics of sheep farm restrictions and the persistence of radionuclides in upland soil soon sank from public view. Thus, the nuclear industry was able to exclude the impact on Britain from its publications on Chernobyl. A telling example is the 1988 Watt Committee report, which I analysed in the previous chapter.[67] Although this report comprised chapters on 'Affected Areas of the UK' and 'Contamination of Foodstuffs', it did not discuss sheep farm restrictions and potential effects of consuming contaminated food. The levels of wet and dry deposition were mentioned, as was the government research, but there were no references to studies on the uptake of radionuclides in British sheep and fish. The same is true for the report's 1991 update, which I will analyse later. Whenever publications in the mid-1990s referred to British sheep farm restrictions, they mostly did so to illustrate the geographical range of the fallout; and this topic seems to have featured more in international than in British publications. The handful of publications dedicated to this topic are essentially the studies by Brian Wynne.

Along with the academic STS-community, mainly the NRPB published on British sheep farm restrictions. In 1998, A.S. Nisbet and R.F.M Woodmann compiled a sixty-page glossy brochure summarizing the monitoring programmes in place and discussing how to improve the current compensation scheme.[68] Yet, this publication considered the issue of radioactivity in sheep a predominantly administrative problem. Therefore, it prioritized the challenge of implementing a more efficient and cost-effective monitoring and compensation scheme, not discussing issues like the dangers of radioactivity in foodstuffs, nuclear power or public health. The authors played out various scenarios for managing the sheep monitoring system, assessing them with the following criteria: 'technical feasibility, capacity, cost, impact and acceptability'.[69] They concluded that 'the extensive nature of the current restrictions in Wales means that none of the alternative options to mark and release monitoring is practicable'.[70] As the current system was too expensive due to the large number of animals being monitored, the authors thought there was only one solution: 'Priority should be given to reducing the size of the restricted area. It is expected that a comprehensive derestriction survey would identify only a few tens of farms by the use of temporary stockproof boundaries'.[71] Over the following years, this recommendation was indeed implemented[72] and the number of sheep farms subject to the restrictions was gradually reduced.

But the government's handling of sheep farm restrictions had been criticized much earlier than the late 1990s, notably outside the NRPB. Already in 1986, heated arguments began between the public authorities and scientists versus the sheep farmers and landowners about issues with the initial implementation. Soon the criticism shifted to the effectiveness of the restrictions and the underlying policies. The government was accused of using sheep farm restrictions as a way to boost public trust in British foodstuffs – the measures were considered a public relations strategy for which the sheep farmers had to foot the bill. An article in the *Guardian* in June 1990 summed up this criticism neatly:

> The most embarrassing thing for the Government is that it is stuck indefinitely with a system of compensating farmers for radioactivity in their sheep which was attractive as a short time measure, but is beginning to look increasingly expensive and questionable. Many scientists believe the restrictions are unnecessary, but politicians fear that removing them could produce a consumer outcry and jeopardize exports. This means that the Ministry of Agriculture is sticking with the scientists and defying public opinion over 'mad cow' disease, but disregarding scientists and bowing to public concern over radioactive sheep, even though this has cost the tax payer £7 million so far, paid mostly in North Wales.[73]

This quote not only illustrates the critique levelled at the protraction of the sheep farm restrictions, it underpins my hypothesis in the previous chapter: scientists in Britain, unlike their French counterparts, were not considered part of the nuclear techno-political regime. They were perceived to be independent from the government; and the government was perceived to have used them as pawns in the pursuit of its own goals.[74]

The entire debate over sheep farm regulations may have seemed a perfect platform for British anti-nuclear groups to use in their anti-nuclear campaign. After all, the number of sheep farms subject to the restrictions only decreased due to more accurate identification and localization of hotspots, not due to any reduction in the level of dangerous radionuclides in the soil. Moreover, the contaminated soil was located in Brits' most loved hiking destinations: the Lake District and Snowdonia. Anti-nuclear activists could very well have referred to these hotspots as reminders of what can happen when a severe nuclear power accident occurs even thousands of kilometres away, thus drawing attention to the danger of nuclear facilities within a few miles of the Lake District and Snowdonia. Yet, such a strategy was not adopted by anti-nuclear groups. What is more, in contrast to France, these persistent hotspots were neither considered a threat to public health, nor did they spark a debate about potential health effects of the fallout. This difference may be because studies on the contamination of highland soil revealed that the radioactivity levels of 1986 were only partially caused by Chernobyl. On one farm in Cumbria, on which the *Guardian* reported in 1990, '35 per cent of the caesium came from atmospheric nuclear weapons testing, and 15 per cent from routine discharges from the nearby Sellafield reprocessing plant'.[75] Such reporting probably lowered the public perception of the severity of the health threat from Chernobyl. At the same time, this re-evaluation of the fallout discursively underpinned the anti-weapon and anti-Sellafield argument rather than affirming the Chernobyl fallout in Britain as an independent anti-nuclear argument. Therefore, public interest in the sheep farm restrictions almost entirely disappeared after the mid-1990s. It was not impossible, however, for people to know about the lasting legacy of the Chernobyl fallout in Britain. Although newspaper reports were rare, some did appear, mostly in the days surrounding the anniversary of the accident. But the protraction of the restrictions on sheep farms was not considered a major news event. None of the articles made it to the front page, and in some years in the 1990s and early 2000s, no articles on the sheep farm restrictions were published in the *Guardian* and the *Observer*.[76] The few articles that were printed described the scale of the restrictions and showed that at least some sheep farmers, whose statements formed an essential part of the articles, were willing to talk about this issue.[77] Apparently, the public did not demand more information or clarification, and the little information that was reported was not linked to doubts or questions about the national nuclear fleet or criticism of the British radiation protection authorities.

The situation in France was quite the opposite. Media reporting on persistent fallout hotspots in France nearly always referred to the *affaire Tchernobyl*. Thus, the material legacy of the Chernobyl fallout in France was discursively embedded in the criticism of the French techno-political regime, particularly when the credibility of the radiation protection authorities was questioned. The moment they proclaimed in 1986 that the fallout would have no negative impact in France, radiation protection authorities became central actors in the French debate on Chernobyl health effects. The research they conducted in the following years was, not least, intended to prove that their initial evaluation had been correct. For instance, from 1996, the IPSN annually published a dossier on state-of-the-art Chernobyl research; the IPSN never failed to assert in the conclusions that the whole debate about observable health effects in France was pointless. Yet, every comment released by state experts regarding health effects only fuelled the debate by giving critics a new opportunity to lambaste the official evaluation.

The debate on the health effects in France already surfaced in the second week of May 1986, hand in hand with the *affaire Tchernobyl*. The logic underpinning the critics' position was as follows: if French public authorities had lied about the true levels of fallout that hit France, then the revelation that the fallout had been even more intense would consequently imply a greater threat to public health than the experts had led the public to believe. However, shortly after May 1986, this debate lost its momentum. Fodder for the debate was removed when the public authorities wrapped up their investigation and analysis of Chernobyl. Critical voices like CRIIRAD continued to call for a re-evaluation of the fallout intensity, but this did not happen; the public authorities had effectively blocked further criticism by eliminating the opportunities for critics to dispute the official evaluations. The French nuclear industry also adjusted its communication activities, no longer stressing the nuclear origin of France's electricity supply. Thus, the whole topic of nuclear energy became less visible in the public sphere in the late 1980s, making it more difficult for critical voices to mobilize and rally a broader audience. Antinuclear activists and critics directed their efforts against incidents in French nuclear power plants as well as the strongly contested construction of the fast breeder reactor Superphénix. As in many other European countries, French public institutions drafted a series of reports on Chernobyl for the national government and international organizations such as the IAEA. But these reports barely attracted any attention outside academic and political circles, and did not incite the production of counter literature.

This changed in the early 1990s and especially in 1996, when the issue of the fallout on France was re-discovered by a broader public. On the tenth anniversary of Chernobyl, the public radiation protection agency IPSN published a detailed brochure, primarily for journalists.[78] The nuclear experts repeated their evaluation of the situation in 1986, declaring that, after ten years, there was still no evidence that the French government and scientific elite had made any errors.[79] This brochure did not have the desired effect. In their reporting on the Chernobyl anniversary, journalists did not praise French clear-sightedness, but reminded their readers of the *affaire Tchernobyl*, pointing to the IPSN brochure as evidence that the contested official narrative had still not changed. Another event at the end of 1996 decisively launched Chernobyl from the past directly into the present: in the Vosges Mountains, a hunter shot a wild boar, which when subjected to a random veterinary analysis, proved to be significantly contaminated. Tests and measurements in the surrounding forest revealed that mushrooms and berries had far higher rates of radioactivity than normal. Thus, in 1996, the legacy of the Chernobyl fallout on French soil made its public re-appearance. The boar incident sparked a search for other hotspots. Measurements were taken in areas of high rainfall in 1986, and CRIIRAD revealed that hotspots were not just in the Vosges, but also in the Mercantour (close to the Italian border) and in Corsica. Concurrent with this search for the still-present radionuclides in French soil, was an active search for the French victims of this radiation, focussing on the claims by doctors in Corsica of increased thyroid diseases in children after the Chernobyl plume had passed over Europe.

In an attempt to end the speculations about the 'true impact' of Chernobyl in France, the Direction de la sûreté des installations nucléaires (DSIN) and the Direction générale de la santé (DGS) commissioned the IPSN and the OPRI – the successor to the SCPRI – to provide a synthesis of all the information currently available on the effects of Chernobyl in France.[80] In November 1997, the IPSN and OPRI delivered their report.[81] By compiling the various groupings of data from the various agencies and institutions, the IPSN and OPRI were able to generate a map of France indicating the total fallout deposition. For some regions, this map was significantly different from what the SCPRI had presented in 1986. The authors of the 1997 IPSN-OPRI report emphasized that the levels of radioactive contamination of foodstuffs had exceeded the 1986 safety levels; also, that there were constellations where the contamination levels of foodstuffs, especially mushrooms, might still exceeded the defined safety limits. The report contained extreme case scenarios on the specific diets of people living in the particularly exposed regions of the Vosges, Mercantour and Corsica. The kind of health effects these doses might have had on the population, however, were not discussed. Like their British colleagues, the French scientists remained rigorously within the parameters of their topic: their task was to conduct a radio-ecological and dosimetric analysis of the fallout, not a study on low-level radiation health effects.

When the report was published as a book two years later,[82] the authors, aware of the pressure from the French Chernobyl debate, felt compelled to comment on the health-related effects of the fallout. In their conclusion, they stated that it might indeed be justified to commission an epidemiological study of thyroid cancers in children in some specific regions of France, not because the estimated doses might have observable negative health effects, but rather as a clear response to the many inquiries from the public and medical practitioners.[83] The foreword by the DSIN's director, André-Claude Lacoste, was in line with the conclusion. He wrote: 'I have walked away with several important lessons from this book: with regard to the estimated doses, the influence of the fallout from the Chernobyl catastrophe in France has remained below a level that could have provoked a justified reaction to a health plan'.[84]

But the critics of the official narrative on French personal affectedness, in particular CRIIRAD and numerous people suffering from thyroid diseases, certainly did not consider that this study proved no observable health effects in France. The IPSN's publications were rather proof of the French *nucléocratie* still trying to cover up their lies of 1986. In response to these tenacious accusations, the DGS commissioned the IPSN and the Institut de veille sanitaire (InVS) to compile a report on the health effects of the Chernobyl fallout in

France. They delivered this report in December 2000.[85] However, no epidemiological study was carried out: the report results were derived from model calculations based on the IPSN's 1997 report on the radio-ecological and dosimetric consequences of the fallout in France. The authors stated right away that such calculations were very problematic given the lack of knowledge on the health effects of exposure to low-level radiation, particularly internal exposure. Despite this admission, the report concludes that an increase in thyroid cancers in French children provoked by Chernobyl would not be observable. Even if individual cases existed, they could not be discerned from the total amount of naturally occurring thyroid cancers.[86] The report readily admitted a measurable and observable increase in thyroid cancers in adults, but stated that this increase had been observed at a global, not only a national level. Moreover, because this increase was already manifest prior to the accident, its occurrence could not be attributed to the Chernobyl fallout. No explanation had been found thus far for the increase.[87]

This evaluation did not convince CRIIRAD and the many French citizens afflicted by thyroid diseases. This was why the Association française des malades de la thyroïde (AFMT) together with CRIIRAD filed a suit in 2001 at the Tribunal de grande instance of Paris.[88] The claim was: 'failure to protect the French people in general and groups of risk in particular against the radioactive fallout of the Chernobyl accident'.[89] Their claim was supported by numerous single claims and later supplemented by class action suits, bringing the total number of complainants involved in this case to some 500 in 2006.[90] This claim and other individual claims were brought primarily against Pierre Pellerin but also members of the 1986 French government.[91] In response to the severe public criticism of Pierre Pellerin, many professionals in the French nuclear sector openly took his side. Thus, with the advent of the 2000s, the debate about the health effects in France became entangled in judicial proceedings and public side-taking. The issue of the experts' credibility and responsibility had become one of personal accountability – and monetary compensation.

Nuclear Power Industry

After EDF published its 1986 brochure, the company did not communicate about Chernobyl to a broader public. One important reason may have been that the EDF did not want to unintentionally create any links between its own activities and the accident. Generally speaking, the company did not place particular emphasis on the source of French electricity in the late 1980s. This changed in the early 1990s, when nuclear power became the focal point of the EDF's advertising and PR campaigns.[92] In 1992, the EDF issued an information brochure on Chernobyl for its employees.[93] This brochure provides a very clear narrative of Chernobyl, consisting mainly of classifying information published in media reports as 'true', 'false' or 'uncertain'. One piece of information classified as true was that the accident was caused by human failure and the faulty design of the reactor, the RBMK, which was different from French plants.[94] The statement that child mortality had increased in the most contaminated areas of Eastern Europe was labelled false; on the contrary, the brochure stated that the infant mortality rate had actually gone down thanks to improved medical surveillance.[95] The claim that there was an increase in cancer rates due to radiation from Chernobyl was also classified as false. The observed increase in thyroid cancer in Belarus was only due to an increase in cancer screening.[96] In general, the causal link between the many illnesses in people living in the most contaminated areas and the radioactive fallout was considered false. These illnesses were actually 'linked to stress, the modification of daily habits and the poor quality of nutrition'.[97] The narrative in this EDF brochure clearly drew on the concept of radiophobia. It also played on anti-East European/anti-Soviet stereotypes, by asserting that the (Western) medical aid brought to the most contaminated regions after the Chernobyl accident was far more advanced than what the Soviet health system could provide. Regarding personal affectedness, the EDF took a clear and decisive stance: any information stating that the public authorities had denied the presence of fallout on French soil was false. The brochure asserted that the opposite was true, that on 30 April 1986, the SCPRI had sent a communiqué to the press agencies stating that the radioactive plume had reached France.[98] The EDF brochure did not just discuss, but also took an active stance on the health effects of the radioactive fallout in France, asserting that it was false that Corsica was one of the most contaminated areas in the whole of Europe; and thyroid cancer rates on the island were not abnormally high.[99] Once it had made these points clear to its employees, the EDF published no further statements on Chernobyl. It continued to systematically distance itself and its activities from the accident.

Simultaneously, in Britain, on the fifth anniversary of the accident, the Watt Committee's Chernobyl working group met again to review the material published on Chernobyl in the meantime and to present a compendium of the reports they judged trustworthy.[100] The Committee stated its motives: 'the material published by the popular media often fail[ed] to separate clearly opinion from fact or to give due weight to the important as compared to the trivial'. Thus, the Watt Committee intended to take on the role of independent expert in the Chernobyl debate once more so that it could inform the public in a purely scientific and neutral manner what information was true or false. The foreword gives us an interesting insight into what information the authors considered unbiased:

> It is the Watt Committee's experience that, at least in the United Kingdom, technical experts in the nuclear industry present frank and objective reports, certainly as far as publishing data. The IAEA presents material which is reliable although it is an organisation which publicises the world nuclear scene as well as regulating and inspecting installations. The British official bodies such as the National Radiological Protection Board also have no interest in promoting the nuclear case.

The Watt Committee's account of how the accident unfolded, the release of radioactive particles into the atmosphere and the effects on health and the environment was thus based on reports, papers and press releases published by the IAEA, the NRPB, the UKAEA and other official institutions as well as articles published in scientific journals such as *Health Physics* and *Radiation and Health*.[101] To a large extent, the interpretation of the Chernobyl accident and the narrative in the report were identical to those in its 1988 publication. The death toll remained unchanged at thirty-one. Yet, the denomination 'volunteers' had now been put in inverted commas, indirectly implying that the Soviet crisis management of the accident had been less than democratic. However, the Committee continued to maintain that there were no health effects in this group. Although they stated that 'no special checks on the health of those engaged in the clean-up operations have been published in the review material', some sentences further down they specified: 'there is no reason to anticipate any radiation linked health problems with the clean-up workers as long as the published irradiation doses are realistic'.[102] Regarding health effects in evacuees and people living in the most affected regions, the authors not only gave an all-clear – essentially dismissing any negative health effects on these groups – but declared that the health situation in children had even improved. Under the section title 'Birth Defects and Increase in Birth Mortality' they cited a representative of the Kurchatov Institute in Moscow: 'growth in infant mortality rate has not been observed in the contaminated regions since 1986 and in most cases has been reduced'. Furthermore, 'in Gomel, infant mortality decreased from 16 per thousand in 1985 to 12 per thousand in 1988'.[103] On leukaemia and other cancers, the authors' comments were brief, because, as they explained, 'there are no reports at this stage, or data, on cancer incidence in the area'.[104] The Committee did include more statistical information on thyroid cancer, asserting that the seventeen thyroid cancer cases verified between 1986 and mid-1989 were 'in reasonable agreement with the prediction'.[105] Despite the lack of detailed data, the authors affirmed that 'there is no reliable evidence that harmful health effects had been observed by June 1990'.[106] The Committee's statement regarding Western and Southern Europe conformed with its evaluation of the health situation in Eastern Europe: 'The medical effects of the fallout over Western and Southern Europe must be small and undetectable'.[107]

Therefore, there were no particular discrepancies between this publication and the narrative in the 1988 report. However, one new element was included: the authors added an eloquent description of the radiophobia concept. By this time, radiophobia had already been renamed and now figured under the expanded and more politically correct phrase 'post-Chernobyl stress-induced illnesses'. The word 'radiophobia' only appeared in the last paragraph of the conclusion, but the concept was paraphrased throughout the text, for instance in an extract from a 1991 Red Cross Survey, that 'psychological stress and anxiety was evident in the Red Cross team who found that people were anxious to know their radiation dose and about future pregnancies'.[108] Within the context of a 1989 IAEA report, the Watt Committee similarly quoted Morris Rosen (Head of IAEA's Department of Nuclear Safety): 'The health effects reported as the outcome of Chernobyl may be due to factors other than radiation … : deficient diet, increase in medical examination, better diagnostic technique and added stress and anxiety arising from the current uncertainties'.[109] In their conclusion, the authors of the Watt Committee report gave a name to the interpretative frame they had applied throughout their assessment of the health situation in the most affected regions and presented their recommendation for concrete action. They ended the report with:

> While the radiation linked to illnesses is likely to be low because of the timely precautions taken by the authorities, these same precautions, and the lack of credible information, have led to a serious decline in the general health of the population, at one time called 'radiophobia'. … The way that radiation is dealt with by the media, the obscure terminology and the links with cancer has led to an exaggerated fear of small doses of radiation. One of the lessons of Chernobyl is that it is urgent to present powerfully in an easy to understand form a balanced view on the health impact of low-level radiation.[110]

The Watt Committee report also described the impact of the Chernobyl fallout in Britain. The authors cited the amount of fallout that had passed over and settled on British soil, stating that 'while in Southern England, where there was little rain while the plume was overhead, radiation levels were 100 – 1,000 Bq/m^2, wet deposition of Caesium 137 in Cumbria and South of Scotland gave levels from 10,000 to 25,000 Bq/m^2'.[111] A map was inserted showing the wet and dry depositions of Caesium 137 in the UK. However, the authors made no mention of potential health effects of these radionuclides in humans, not even in the chapters on 'Environmental Effects' and 'Farms'. Thus, while classifying the debate on Chernobyl health effects in Eastern Europe as radiophobic, the Watt Committee report provided a clear statement against health-related personal affectedness in Britain.

Critical Voices

The GSIEN activists had been among the first to contest the official French statements on Chernobyl's impact in 1986, in particular Bella and Roger Belbéoch. In the 1980s, the Belbéochs were among the best known and most publicly visible French anti-nuclear activists and also co-founders of the Comité Stop Nogent in 1987. In May 1986, Bella Belbéoch wrote an article for *Ecologie*, the Société française d'écologie magazine. Her predictions for the trajectory of the official Chernobyl evaluations became a reference point for critical voices in the French Chernobyl debate. On 1 May 1986, she wrote:

> In the coming days, we can expect an international complot of official experts who will try to minimize to the maximum the number of victims of this catastrophe. The continuance of the civil and military nuclear programmes imposes on the collectivity of states a tactical complicity that exceeds ideological or economic conflicts. The international health organizations, which are in principle independent from the states but strictly controlled by the Great Powers, will serve as liaison agencies to uphold the appearance of objectivity and neutrality.[112]

Because of their concerns, Bella and Roger Belbéoch continued to closely follow the work and reports of the international committees and expert groups put in charge of evaluating Chernobyl's impact in the most affected regions of Eastern Europe. They scrutinized the publications released by the WHO and the IAEA. In 1992, in their contribution to the first edition of the journal *L'Intranquille*, they addressed the wider public, strongly criticizing the official narrative diffused internationally about Chernobyl's impact.[113] The following year, a revised and expanded version of this article was published as a book.[114] In it, the Belbéochs discussed various aspects of the Chernobyl issue: the contemporary situation in the most affected regions, the struggle to define the dose limit for evacuation, the changing estimates of the number of victims, the alienation strategies applied in the West and so on. A key aspect of their argument was contesting the radiophobia concept that featured prominently in the Soviet and international expert groups' evaluation. In the chapter 'Le complot international' (the international conspiracy) the authors incorporated all the aspects on Chernobyl they had presented into one claim: 'There is nothing surprising about what is happening at the moment within the circle of experts'.[115] As Bella Belbéoch had predicted in her article for *Ecologie* in 1986, all Chernobyl-related activities and communications were aimed at protecting the international civil and military enterprise from being profoundly questioned:

> The complicity of the Western experts, scientists, technicians, doctors, sociologists and specialists in humanitarian aid was, without reservations, aimed at helping the central powers to 'manage' the social, economic, and political crisis that had emerged as a consequence of the accident. It was necessary to convince the people living in the contaminated areas that they had nothing to fear about their health. Those who were sent to 'liquidate' the consequences of the catastrophe at the site itself were not to question the doses of radiation they received or would receive. The rapid restart of the undamaged reactors provided the proof that there was nothing to fear from nuclear energy.[116]

Although the book focussed on the (health) impact of Chernobyl in Eastern Europe and did not tackle the radioactive fallout in France, the impact of Chernobyl on the French nuclear debate was indeed addressed:

> For a long time our experts were the primary providers, if not the producers, of information. They simply denied the catastrophe, reduced the event to some 30 victims. But it was impossible to conceal the mass evacuations, and the non-return of the people to their homes was clear proof the soil decontamination had failed utterly. Because the reality of the catastrophe could not be denied, the discourses changed: Chernobyl had indeed been a major catastrophe, but its objective consequences were minor.[117]

Like the early critical voices in Britain, Roger and Bella Belbéoch directed their criticism against the global nuclear policies at stake in the official Chernobyl narrative. The difference was, that while the aspect of possible health effects from low-level radiation exposure was central to their analysis, this issue had barely been discussed in British publications and was usually referred to as just an unresolved issue.

More articles on Chernobyl appeared in GSIEN's magazine *La Gazette Nucléaire*, criticizing the official Chernobyl narratives provided by international expert groups. At the same time, Soviet dissident voices became available to French readers. In 1990, Grigori Medvedev's book on the causes of the accident was published in France,[118] and Alla Yaroshinskaya's book followed in 1993.[119] These examples illustrate that, in the early 1990s, the focus of the French Chernobyl debate was on revealing the 'truth' about Chernobyl with regard to Eastern Europe. Only after the tenth anniversary of the accident did public interest switch to the 'true' impact of Chernobyl in France.

It was the CRIIRAD who led the struggle to reveal the 'true' level of contamination in France. With this aim in mind, Michèle Rivasi founded the organization in May 1986. The 1997 report released by the IPSN on the impact of Chernobyl in France (discussed above) proved to Michèle Rivasi and the CRIIRAD, that nothing had changed since the Chernobyl accident – the French radiation protection agencies were still not to be trusted. A book that Michèle Rivasi published together with Hélène Crié in 1998 underlined this opinion: 'Twelve years later, in defiance of some attempts to appear more "transparent", the nuclear authorities have not changed how they function: disinformation and incompetence remain the rule.'[120] Thus, the only way to reveal the truth was to carry out independent studies. And this is precisely what the CRIIRAD had been doing for years in south-east France and Corsica. To complement these selective studies, the CRIIRAD commissioned geologist André Paris, in 1999, to take soil samples all over France. In 2002, the results of CRIIRAD's measurements and analyses were published in the book *Contaminations radioactives: atlas France et Europe*[121] (Radioactive contaminations: atlas France and Europe). The editors' aim was clear: 'This atlas of reference provides unpublished maps and information, which are useful for anybody who wants to understand the debate on the sanitary consequences of the Chernobyl catastrophe.'[122] The opening section of the book was written by CRIIRAD's director, Corinne Castanier. She described the 1986 *affaire Tchernobyl* and the history of CRIIRAD, before providing a comparison of the public authorities' official figures for radioactive contamination in France with the results of the CRIIRAD studies. The subtitle to her chapter said it all: 'Les preuves du mensonge' (the proofs of the lie). According to the CRIIRAD, the biggest and most serious lie was SCPRI's declaration in early May 1986 that the readings needed to be 10,000 to 100,000 times higher before significant problems in public health would begin to occur. For CRIIRAD, the atlas proved that the readings had indeed been high enough to warrant safety measures, and by taking an uncompromising position against the adoption of any safety measures in early May 1986, the French public authorities had wilfully exposed the public to a health threat.

The claim that French public authorities had knowingly exposed people to a health threat – specifically one targeting the thyroid gland – was met with marked interest by those suffering from thyroid diseases. Because the thyroid gland is very receptive to (radioactive) iodine, exposure to this radionuclide can cause severe illnesses and even cancer in this organ, especially in people exposed at a young age.

In 1999, some French patients founded the AFMT. That same year, at the organization's general meeting, the members were offered a concise explanation for their illnesses: Jean-Michel Jacquemin presented them his book *Ce fameux nuage... Tchernobyl, la France contaminée* (The famous cloud… Chernobyl, France contaminated), in which he proclaimed his theory that the increase in thyroid diseases in France was a result of the Chernobyl fallout. I will discuss Jean-Michel Jacquemin, his publications and his role in the French Chernobyl debate later on. Significant in relation to the AFMT, is that Jacquemin was able to convince many AFMT members of the veracity of his theory: the support group partly transformed into a group of complainants, which two years later and together with the CRIIARD filed a suit in the French courts.

AFTM activities provoked such severe criticism among French radiation physicians that, in 2005, a group of about fifty doctors published a manifesto in *Libération*.[123] This manifesto addressed 'people with thyroid diseases who ascribe their illness to the passing of the radioactive cloud over France in 1986'[124] and emphatically communicated: 'NO, there is no "Chernobyl effect" in France'.[125] According to this manifesto, the worst effect of the whole debate, was that many people began to challenge or question the expertise of their doctors, thereby putting their health at risk by seeking absurd explanations for their illnesses. The text concluded by identifying who was responsible for what they felt were absurd claims: 'These sick Frenchmen are the hostages of an anti-nuclear and juridical-medical lobby'.[126]

Although from the mid-1990s, the French debate on Chernobyl health effects focussed very much on the effects in France, public interest in Chernobyl's health effects went hand in hand with increasing awareness of the situation in the most affected regions in Eastern Europe. The living conditions of people in the most affected regions became a key topic of interest to a group of sociologists from the University of Caen (Normandy). The 'Caen-Group' is not a self-coined term, but one I use here to refer to the people involved in Guillaume Grandazzi and Frédérick Lemarchand's publications, looking at Chernobyl from a socio-philosophical perspective. Grandazzi and Lemarchand are the editors of the anthology *Les silences de Tchernobyl* (The silences of Chernobyl), published in 2004.[127] The second edition of this book along with Grandazzi and Lemarchand's work and views were widely referred to in the French media reporting on the twentieth anniversary of Chernobyl. The key message in Grandazzi and Lemarchand's interpretation was that they did not see Chernobyl as an isolated solitary event that took place in 1986 – but an ongoing event that, though it began in 1986, will continue to unfold in the future.[128] In line with Günther Anders and his considerations on the atomic age, Grandazzi and Lemarchand saw in Chernobyl the manifestation of an era characterized by the constant possibility of total destruction. Furthermore, they believed that time was inverted in this era, when the past becomes the future. And therefore, what was needed was a completely new form of commemoration, one that commemorated the future.[129] Because Chernobyl broke with all temporal and spatial reference points, the problem arose, how do you create an image of the unimaginable? Another key element regarding the lack of reference points is the missing *événement fondateur* (inceptive event). The victims' daily lives suddenly changed without an evident reason: 'Still today, millions of inhabitants of the contaminated zones find themselves denied a reference point for the accident. … The event is first and foremost daily life and the fact of being brutally thrown into a world marked by new rules, of new interdictions'.[130] Thus, the main objective of *Les silences de Tchernobyl* was to expose daily life in the contaminated areas, and how people's suffering was denied by official institutions and the nuclear lobby. The topics in this anthology ranged from the politics of the Soviet cover-up in 1986 to the current working conditions for doctors in the most affected areas, exemplified by the case of Yuri Bandazhevsky.[131] From Grandazzi and Lemarchand's point of view, one of the main issues was that the people living in these areas were confronted with something hereto completely unknown; there was a total 'absence of reference points, of experiences, of cultural references that could be drawn on in this entirely new situation'.[132] This setting made it difficult for people to improve their own situation and even express their feelings. What worsened the situation for these people was the concept of radiophobia, which was used in official reports to describe their problem. This term transformed the victims into culprits, thus denying their fears, suspicions and consequent caution regarding their environment – their only weapons against the further uptake of more radionuclides and crucial for their survival in the most contaminated areas.

Significantly, the editors of *Les silences de Tchernobyl* dedicated several sections of their volume to the book *La supplication: Tchernobyl, chroniques du monde après l'Apocalypse* (The supplication: Chernobyl, chronicles of the world after the apocalypse) by Svetlana Alexievich. Alexievich's book had been translated into French in 1998 by Galia Ackerman[133] and enjoyed tremendous success: from numerous reprints, to adaptations as plays, to countless quotations in almost everything written about Chernobyl in France, *La supplication* was the incarnation of the counter narrative to the official narrative provided by the French public authorities and international expert groups.

The official national and international narratives on Chernobyl also excited criticism in Britain. Compared to the variety of publications and sheer number of critical voices in France, however, the critical Chernobyl voices in Britain were almost unheard between 1989 and 2005. This applies even more to the aspect of personal affectedness. In general, anti-nuclear power opinions, never as strong in Britain as those on anti-nuclear weapons, steadily lost momentum. To understand the lack of critical voices, we must consider this phenomenon in relation to the changes in nuclear policies: in the 1990s, the British government decided to stop building any new nuclear power plants. The new Sizewell B reactor was only completed and connected to the grid in 1995. But rather than make this nuclear reactor the pioneer of a whole fleet of PWRs on British soil – as originally intended – the government decided in May 1995 that the civil nuclear sector would no longer receive public funding. Consequently, CEGB renounced its plans to build more new nuclear power plants. This energy policy was the opposite extreme of France's nuclear policy. There, the government had decided to replace old PWRs with a fleet of new third generation PWRs, the so called European Pressurized Reactor (EPR), which a joint French-German technical consortium had begun to develop in the early 1990s. Concurrently, France profoundly reorganized its nuclear sector. The government overhauled the radiation protection authorities and founded new agencies, including the Autorité de sûreté nucléaire (ASN) in 2006. But, more importantly for the nuclear industry, a holding company Areva[134] was formed with the aim of becoming a world leader in reactor construction, uranium extraction and fuel production. This policy to give strong political and financial support to the French nuclear power industry incited fierce criticism from anti-nuclear activists, who consequently intensified their protests and campaigns. In 1997, the Réseau sortir du nucléaire (network phase-out of nuclear power) was created and became a key player in coordinating these campaigns. So, while in France the anti-nuclear power protests made a strong comeback in the late 1990s, in Britain there was barely any such activity. Since phasing out existing plants had never been an objective of most anti-nuclear power campaigners, their case was basically won with the government's decision of 1995, of course with the notorious exception of Sellafield, which continued discharging into the Irish Sea. If anti-nuclear power plant activism was almost non-existent in the 1990s, activism associated with other environmental issues was thriving in Britain at the time. In the early 1990s, Greenpeace and FoE membership soared. In particular, Thatcher's plans for new road construction and large-scale projects like the Eurostar had incited intense protest, which was strongest in south-east England.[135]

In the late 1980s, FoE was still actively involved in the debate over the impact of the Chernobyl fallout, in particular the fallout in Britain. In a 1989 report *Fallout over Chernobyl: A Review of the Official Radiation Monitoring Programme in the UK*, FoE campaigners P. Green and P. Daly criticized how the sheep farm restrictions had been and were being managed and claimed that in some areas the restrictions had been lifted too early, whilst in some affected areas they had never even been introduced.[136] The report did not pass unnoticed, as may be attested by its mention in an article in the *Guardian*:

> A group of scientists yesterday added their voices to those of Friends of the Earth in suggesting that the ban on the sale and slaughter of sheep introduced in parts of Britain after Chernobyl may have been wrongly applied. The scientists say results obtained in north-west Wales show that farmers in some affected areas may have wrongly escaped the ban and others may have been unnecessarily restricted.[137]

In 1991, FoE released another publication on Chernobyl, this time addressing the situation in the most affected regions of Eastern Europe. This publication was a six-page brief written by FoE radiation campaigner Patrick Green after his visit to Belarus. Green exclusively discussed the health effects in Belarus and was very cautious in how he crafted his statements. This is evident from his comment on the observed increase in illnesses in these regions: 'However, these statistics only provide a general indicator of a medical trend, they do not prove whether or not the exposure to radioactivity was responsible'.[138] In spite of this prudence, Green criticized straightforwardly the IAEA's evaluation of the health situation: 'It is extremely worrying that the international agency charged with investigating the health effects of the accident seems to have made up its mind that the only health effect is radiophobia'.[139] He indicated the need for epidemiological surveys, but from Green's point of view, whether these studies should be conducted by the IAEA and whether or not the right people would be included, remained to be seen. I found no further material on Chernobyl published after the early 1990s by either FoE or other British environmental NGOs.

With regard to British sheep farmers, no further publications seem to have been released on Chernobyl after the 1988 memorandum by the Farmers' Union of Wales. However, given that British newspapers on some occasions reported on the continued sheep farm restrictions and at times wrote longer articles around the individual lives of affected farmers, some public statements are available for this period. Undoubtedly, using these articles as sources involves looking at the sheep farmers' statements through the lens of an intermediary, the journalists; but with no firsthand sources available, these articles nevertheless provide information on the group of people most affected by the Chernobyl fallout in Britain.

In contrast with the FoE's viewpoint that restrictions had been lifted too early or had not even been imposed, some sheep farmers found the existence of the restrictions problematic, and believed they were the victims of a government food safety PR campaign. This perception was clearly expressed in an article in the *Guardian* in June 1990: 'Roger Ward, regional director of the National Farmers Union in Cumbria, says his view all along was that the restrictions weren't necessary, but the NFU went along with them to demonstrate to the consumer that the product was definitely safe'.[140] In addition to the mistrust and frustration with public authorities, the restrictions had proved to be a financial challenge for many farmers: though they received compensation payments, some farmers were extremely worried that they would be unable to find a buyer if they decided to sell their farm in the future. Others were concerned over the possible health effects the fallout would have on their families and their animals. Trebor Roberts, a sheep farmer from North Wales, told a journalist in April 1990:

> Personally I am scared for the young people and the children. I am too old but what about seven and eight-year-olds when it happened? Some people locally were taken for testing and were found to have very high radiation

> levels indeed. I have worked this farm for 21 years and bought it just before Chernobyl. During all that time I had maybe six deformed lambs. Last year I had more than 20. The mothers of these lambs were born after Chernobyl. Other farmers have had similar experiences. ... Nobody seems to care.[141]

Aside from Benbecula, Trebor Roberts' statement is one of the very few instances of openly addressing the perception of an observable sanitary impact of the Chernobyl fallout in Britain. However, the author of this article seems to have found Roberts' statement rather extreme. The article balances Roberts' worries about possible health effects in children in this region with a statement from physicist Dr David Sanderson: 'Since the meat contains only tiny amounts of C137 and is distributed widely by the food industry, he says, only small amounts are actually eaten by any one individual. And the absorption of a little more of an element that is already naturally there is not a major consideration'.[142]

It seems that not too many farmers in the area were willing to give interviews to journalists, as nine years later it was once again Trebor Roberts who appeared in an article published in the *Guardian*. In this 1999 interview, Roberts strongly reiterated his argument from 1990 that 'Nobody seems to care' when he spoke about the widespread ignorance regarding the contamination in Snowdonia: 'I could even go into Dolgellau [the town closest to his farm] and people on the street would wonder what restrictions you were talking about. They have not a clue, even seven miles away'.[143]

While some farmers might have been happy that the enduring sheep farm restrictions were cloaked under a veil of ignorance – after all, the BSE crisis was already threatening to discredit British beef and people did not need reminding of the problem with radioactivity in sheep – others did not see these restrictions as just one food scandal like all the rest. They drew wider conclusions and linked them to their nuclear-political stance, like for instance a farmer from Wales, who stated in a 1994 interview: 'My sister was in CND and I honestly didn't understand what she was going on about. Then this happens. A cloud of radioactive dust is created 2,000 miles away and travels 5,000 miles up over Scandinavia. It keeps going until it hits a mountain in Wales. And eight years later, it's still here'.[144]

Individual Voices

As mentioned, very few publications on Chernobyl appeared in Britain between 1989 and 2005, other than natural science studies on transport mechanisms in the air, water and soil. The object of this mere handful of books was to present a concise history and analysis of the accident and its impact. The best known are by two academic geographers: *Chernobyl: The Long Shadow* (1989) by Chris C. Park,[145] and *Fire in the Rain: The Democratic Consequences of Chernobyl* (1990) by Peter Gould.[146] Chernobyl also became a popular topic for children's non-fiction books.[147] It inspired Piers Paul Read to write *Ablaze*, narrating the accident in the form of an action thriller.[148] As these books were all one-off contributions to the Chernobyl debate, written by authors who did not publish on nuclear issues on other occasions, I do not discuss them in detail. One of the few individual British voices[149] who had offered an interpretation of the event in the late 1980s and continued writing on this subject, was Richard F. Mould.

Similar to the Watt Committee's approach, Richard F. Mould reviewed his Chernobyl account published in 1988 and released it, in 2000, as what he called 'the definitive history of the Chernobyl catastrophe'.[150] The task he set for himself in this book was nothing less than to provide 'a balanced account of the accident and its aftermath, excluding media hype and biased accounts of self-interest groups, and debunking some of the myths which have surrounded Chernobyl'.[151] In the foreword, Mould said he believed he had fulfilled this task. To convince the reader he was qualified to do so, he took an entire page to present his about-the-author information, listing his books, his work for the WHO and IAEA, his honorary memberships and the number of his grandchildren.[152] Unlike in his 1988 publication, the pictures in this new edition illustrated the text instead of the other way round. Furthermore, the language style did not have the elitist haughtiness of his previous book. In general, the text was neither dogmatic nor agitated; it cited many reports and various sources. To ensure a balanced representation of the facts concerning the unfolding of the accident, Mould used a combination of eyewitness accounts, scientific reports and newspaper articles. When expressing his position on nuclear power, he now adopted a more matter-of-fact attitude, stressing that the power plants were there and therefore we just had to deal with them. He no longer praised the 'only viable alternative for the foreseeable future' as he had done in 1988. This change in language corresponds with what Sezin Topçu describes in France after Chernobyl as a shift in the nuclear industry's discourse, from the sacralization of nuclear power to emphasizing the *fait accompli*.[153]

Mould detailed his evaluation of the fallout's health effects in chapter 15. I quote from the introduction to this chapter, firstly because it clarifies the radiophobia concept; secondly, it demonstrates the significance of the Chernobyl narrative within the framework of radiophobia for comparing the accident to other 'technological disasters':

> Technological disasters in the period 1984–86 have included not only that at Chernobyl but also the explosion at the chemical plant in Bhopal, India

> and the crash of the American space shuttle Challenger. In the wake of these, people have grown more sceptical of new technologies and more fearful of familiar technologies around them, particularly when they perceive that there is an impact on health. … There is, though, no doubt that Chernobyl is the greatest psychological disaster of the twentieth century, having had a worldwide impact. In 1986, just after the accident, the major cause of concern was the future expected increase in the incidence of thyroid cancer and leukaemia, particularly in children and adolescents. However, a decade later it became apparent that the magnitude of psychological and social problems, of which radiation phobia is only one aspect, far outweighed that of radio-induced cancers, and that the social and associated economic problems of psychological illness in Ukraine, Belarus and Russia, due to the Chernobyl accident, would be enormous, both now and in the 21st century.[154]

To substantiate his interpretation, Mould referred to studies conducted under the framework of the International Chernobyl Project. The reports by the International Chernobyl Project (and its successor the Chernobyl Forum) widely diffused the concept of stress-induced illnesses as the common narrative of Chernobyl's health impact, thus emphazising the psychological problems. Its 1991 report stated: 'there were many important psychological problems of anxiety and stress related to the Chernobyl accident and in the areas studied under the Project these were wholly disproportionate to the biological significance of the radioactive contamination'.[155] The physical condition of the people in this area was considered less problematic as this quote reveals: 'General Health: The children who were examined were found to be generally healthy. … No statistically significant evidence was found of an increase in incidence of foetal anomalies as a result of radiation exposure'.[156] The International Chernobyl Project thus concluded that the health-related problems did not stem from the radiation levels, but 'action should be taken on adult hypertension and dental hygiene as major health issues'.[157] Another important source to corroborate his arguments were the proceedings of the WHO conference in 1995, which Mould cited extensively, for instance:

> The national health registries of Belarus, the Ukraine and Russia recorded a significant increase in many diseases that are not related to radiation. These have included endocrine diseases, mental disorders and diseases of the nervous system, sensory organs, and digestive and gastrointestinal systems. Congenital abnormalities have also been observed. While present evidence does not suggest that these diseases are radiation induced, it is possible that such problems resulted from the considerable stress caused by the accident.[158]

The entire debate on the health impact of the Chernobyl fallout can, putting it bluntly, boil down to this statement, or more precisely, to whether or not you believe it. If you believe it, you need to refer to the radiophobia meta-theory to explain the increase in observed illnesses. Refuting this explanation mobilizes other theories concerning 'alternative' models in radiation biology and the power relations in international nuclear politics. One theory in particular has garnered progressively more support since the early 2000s. It refers to an agreement in 1959 between the IAEA and the WHO, whereby these two international agencies worked out the terms of their relationship, namely: 'Whenever either organization proposes to initiate a program or activity on a subject in which the other organization has or may have a substantial interest, the first party shall consult the other with a view to adjusting the matter by mutual agreement'.[159] This agreement lately sparked an intense debate over the impartiality of the WHO concerning nuclear issues and became the anti-nuclears' major argument for questioning the official number of Chernobyl victims provided by expert groups like the International Chernobyl Project. It produced a movement calling for a 'free and independent WHO': a WHO independent of the nuclear lobby thus capable of revealing to the world the 'truth' about Chernobyl which, in the activists' opinion, the IAEA has succeeded in covering up. Ever since the April 2007 Chernobyl anniversary, the group Independent WHO has organized a permanent vigil at the WHO headquarters in Geneva to raise awareness amongst people working for and visiting the WHO.[160] Thanks to the important role many French anti-nuclear activists have played in the work of Independent WHO, the IAEA–WHO agreement has gained much visibility in the French Chernobyl debate.[161] Even pro-nuclear activists feel obliged to refer to it in their Chernobyl accounts – albeit to dismiss how anti-nuclear activists frame this agreement, classifying their claims as an unfounded conspiracy theory.[162] Like accounts such as Mould's which relied on WHO evaluations, were thus fertile terrain for critical attacks from anti-nuclear activists.

To what extent Mould himself believed there was no connection between the increase in various illnesses in the most affected areas and the radiation levels became even more obvious in his chapter on 'Other Non-Malignant Diseases and Conditions'. Here, Mould presented a slight reformulation of the WHO account cited above:

> Other diseases and conditions have, of course, been reported, such as those of the cardiovascular system and of the immune system, but no correlations have been established with radiation exposure from the Chernobyl accident. There are so many confounding factors, including stress due to the accident, socioeconomic conditions and an inadequate diet, that it is extremely un-

> likely even with long-term detailed follow-up that it will be possible to prove any significant correlations.[163]

Likewise, Mould was extremely sceptical that links could be determined between various types of cancers and radiation. According to him, only an increase in thyroid cancer could be attributed to the Chernobyl fallout. 'In terms of other cancers in Chernobyl populations, no significant increase in their incidence has been observed'.[164] Probabilistic cancer deaths as a result of Chernobyl had of course been calculated, but because of the high background cancer mortality, this increase was not observable. Using typically technical language, Mould stated: 'For the cohort of 200,000 liquidators the increase in solid cancers is some 2,000 compared to the background of 41,500. This excess would be difficult to detect epidemiologically. This is also true for the cohort of 6,800,000 residents in contaminated territories for whom the predicted excess solid cancers is some 4,600 against a background of 800,000'.[165]

The Watt Committee and Richard Mould examples show that the radiophobia narrative was willingly taken up by those actors who, right from the start, had regarded the health impact of the fallout as limited. The concept of radiophobia enabled these actors to uphold this evaluation and highlight the risks of cultivating an exaggerated fear of radiation. Consequently, the issue of British personal affectedness did not feature in these publications, unlike in France, where the topic of personal affectedness was omnipresent.

While the critical individual voices that wrote about Chernobyl in the direct aftermath of the accident did not continue to play a part in the public debate, a new actor, Christopher Busby, rose to partly fill this gap. From the mid-1990s, Busby was the main person publicly speaking and writing on the health-related effects of Chernobyl in Britain. An active member of the Green Party, Busby dedicated his life to finding proof of the health effects that exposure to low-level radiation had caused in Wales. He believed that Wales in particular had been exposed to discharges from nuclear facilities and had received more radioactive fallout from nuclear weapons testing than any other British region.

Busby is a highly controversial figure, and it is almost impossible to find a neutral statement about his work, and even more so about his person. People either consider him a hero fighting the mighty nuclear lobby, or hate him and campaign against him, especially those associated with this lobby.[166] It is remarkable that when writing about Busby, even normally eloquent individuals – like George Monbiot[167] in his blog for the *Guardian* in November 2011 – got carried away, using language you would expect in sensation-seeking tabloids.[168] The talk page behind the Wikipedia article on Busby highlights the controversies: pages and pages of arguments between Wikipedia editors and resolute pro- or anti-nuclear activists.[169]

Busby's first public contribution to the British Chernobyl debate came in the form of his 1995 book *The Wings of Death: Nuclear Pollution and Human Health*,[170] which he published through Green Audit, the environmental consultancy firm he founded in 1992. This book aimed to publicize his 'Second-Event Theory', claiming no less than to have discovered the biological mechanism linking radiation from nuclear power plants to leukaemia in children. The central argument of this theory is that 'internal radiation can have a quantitatively greater effect than external radiation'.[171] This theory, which has not gained ground in the field of radiation biology, is less interesting here than the way Busby integrated the Chernobyl fallout in Britain in his argument.

In *The Wings of Death*, Busby's account of the Chernobyl accident was only one page, and along with Windscale and Kyshtym, came under the headline of 'major accidents in the nuclear industry'. Busby referred to the evaluation by famous US anti-nuclear activist John Gofman, who had calculated that the Chernobyl accident would cause 970,500 cases of cancer. Busby, however, argued this number 'may be still an underestimate' because he believed that the effects of internal radiation had not been correctly considered: 'The Hiroshima survivors data used by him [Gofman] is, itself, flawed in that it did not distinguish internal contamination dose from external dose and made the first of a series of errors which have been repeated ever since in studies which have attempted to evaluate the effects of ionizing radiation on health'.[172] Busby also strongly criticized the government's emergency management in 1986: 'The Chernobyl disaster provided a contemporary insight into the workings of government in Britain. The watchword is always: "deny any risk." So when the radioactive rain fell in Wales and in Cumbria the population was told "No need for alarm – continue as usual."'[173]

Thus, Busby took Chernobyl as an example to illustrate that the existing models used by UNSCEAR and other official institutions were erroneous.

> There are already 450 cancers in the first 10 years for the under-14 age-group alone in the areas into which the evacuees were moved. Only 100 excess thyroid cancers were predicted for all age-groups combined in this population for the next fifty years. … These predictions were made on the basis of the existing risk factors, so their inaccuracy, already apparent and no doubt to become more obvious over the coming years, indicated that the risk-factor calculations for thyroid cancer, like those for leukaemia, are unreliable.[174]

According to Busby, Chernobyl was just one source of radionuclides among many others. But he also asserted that the effects of these radionuclides were far more severe than what official publications had proclaimed, and even in places as far away as Wales. Using data from a study published in 1991 – in which, incidentally, the author declared that Chernobyl had had no effects on infant mortality in the regions with the highest fallout in Britain – Busby came to a different conclusion: 'There was a peak in neonatal deaths at the time of the fallout followed by a statistically significant depression in infant mortality over the nine months following the peak. This depression was followed by a statistically significant excess in neonatal mortality, perinatal mortality, and still-births in the month-long period beginning on 27 February 1987'.[175] From Busby's point of view, there was an obvious explanation for this depression in infant mortality in the nine-month gap: foetal death in utero. And although they may have survived exposure in utero and the first few months after birth, it did not mean these children were healthy. Many showed low birth weights and 'those babies that survived all of these effects may still become sick later, of cancer in childhood or perhaps some mutation-related illness in childhood or later life'.[176] Yet, Busby not only saw effects in foetuses and children and an increase in thyroid cancer and leukaemia, he also noted 'an immediate cancer increase in 1987 across all types of cancer'.[177]

Busby continued his investigations into the health effects from exposure to low-level radiation in Wales and in Britain in general. Chernobyl was only one of many other sources of low-level radiation; he also pointed to Sellafield, various British nuclear power plants and weapons testing fallout. The most attention was probably paid to his publications on childhood leukaemia in infants living in the vicinity of nuclear installations – a highly contested issue that has been investigated in other countries as well. In his studies, Busby treated the health effects of Chernobyl fallout in Britain primarily as a reference point to reveal the entire negative health effects of other nuclear ventures. In 1999, the Low Level Radiation Campaign – which Busby had initiated – published the brochure *Do We Really Want Nine Chernobyl Accidents Every Year for the Next Ten Years?*, which argued against implementing a Euratom-directive on nuclear waste into national law.[178] The brochure presented Chernobyl as one problem among others, emphasizing that 'we all have man-made radionuclides inside us from Sellafield, weapon tests, Chernobyl etc. etc'.[179] As with the general British anti-nuclear discourse, Sellafield figured top of Busby's list. Thus, unlike France, where revealing the 'truth' about Chernobyl manifested itself in the endeavour to identify French Chernobyl victims, revealing the 'truth' about Chernobyl for Busby meant exposing the health impact of the British and the global nuclear enterprise in general.

In 2000, together with Molly Scott Cato (who went on to become a prominent Green politician in the UK and member of the European Parliament), Busby published a paper specifically on Chernobyl: 'Increases in Leukaemia in Infants in Wales and Scotland Following Chernobyl: Evidence for Errors in Statutory Risk Estimates'.[180] In this paper, Busby and Scott Cato claimed that through their calculations, they had discovered 'an error in the presently accepted risk factors for radiation induced leukaemia following this kind of exposure of about 100-fold or more.'[181] After comparing the number of cases predicted by the NRPB to the observed cases, they concluded that 'the effects of the Chernobyl fallout in Wales and Scotland were significant.'[182]

The NRPB took note of Busby's studies, and in the early 2000s published a response dismissing all of his findings.[183] Annex 2 of this document was dedicated entirely to contesting the outcome of Busby and Scott Cato's study on Chernobyl's health effects in Britain. The NRBP concluded that 'the current evidence for increased leukaemia risks associated with exposures from the Chernobyl accident is not convincing overall, and does not support the conclusions drawn by Busby and Scott Cato about radiation risk estimates.'[184] This response only led to Chris Busby presenting the NRPB with another counter argument in which he in turn dismissed the NRPB's arguments against the validity of his findings.[185]

Busby's objective was clear: to reveal the pro-nuclear bias of natural science research, not that of individual scientists, but nuclear science in general. He communicated this aim explicitly in his publications:

> When I began my investigations into the health effects of low-level radiation following the Chernobyl accident, I was guided by a vague suspicion that some of the science supporting nuclear power was unsafe. But during the course of my research I have uncovered so much evidence of cover-ups, deceit, secrecy, and an absence of genuine scientific research that it has become obvious that what most nuclear scientists are about is supporting a paradigm.[186]

Such statements clearly demonstrate why Busby has become such a controversial figure, and why both he and his work have been fiercely criticized by other scientists. This critique and its intensity, however, just seem to have proved to Busby not only that he is right, but that anyone who does not accept his theories must be involved in the nuclear industry's concerted action to cover up the truth regarding the health effects of exposure to low-level radiation.

In France, many more individuals have been involved in the Chernobyl debate than in Britain. And quite a few people from both the pro- and anti-nuclear side have published accounts on the accident, incorporating their personal views on nuclear policies in their analysis of the event and its impact. On the pro-nuclear side, the book by Georges Charpak,[187] a prominent French physicist who won a Nobel Prize, is a particularly interesting contribution to the French Chernobyl debate.[188] His work, arguing strongly in favour of stronger safety principles and nuclear fusion technology to make nuclear energy the main future source of electricity, was very influential: his Nobel Prize and active role as scientist very much in the public eye lent him an extraordinary amount of credibility in the scientific community and with the public at large.

As addressing every important voice in France would make this chapter excessively lengthy, I will only describe one of the most prominent in the debate on French personal affectedness, namely an anti-nuclear proponent who shares some similarities with Busby: Jean-Michel Jacquemin. Admittedly, their publications are not at all comparable on a scientific level. But since the mid-1990s, both men have dedicated their lives to revealing the 'true' impact of the Chernobyl fallout in their respective countries. While physical chemist Busby concentrated on conducting natural science studies, former accountant Jacquemin focussed on bringing justice to French victims.[189]

Jacquemin's public commitment – which decidedly contributed to many people in France linking their thyroid illnesses with the Chernobyl fallout – should be viewed critically. Problematic in Jacquemin's work is the methodology he applied for delivering his statements: he cited material without giving bibliographic references, and his 'cancer index survey' randomly consulted local databases. Another problematic issue was his conviction that he was a martyr on a holy mission.[190] Jacquemin equated the situation in France with the situation in the worst affected areas of Eastern Europe, stretching the comparison to such absurd extremes that even organizations like CRIIRAD distanced themselves from Jacquemin's statements.[191] In 1998, Jacquemin published his first book on Chernobyl: *Ce fameux nuage... Tchernobyl, la France contaminée*.[192] His book discussed in detail the *affaire Tchernobyl*, while the officials and experts' continual 'disinformation' and 'lies' were the leitmotif of the narrative. Jacquemin had set himself the task of revealing the true extent of the contamination in France, which he claimed, had been even greater in some hotspots than in some areas of the restricted zone around the Chernobyl plant. The comparison between radiation levels in France and around the Chernobyl plant featured prominently on the book's back cover.

Jacquemin was neither the first nor the only person to have investigated the 'true' impact of Chernobyl in France: CRIIRAD had already dedicated itself to this task for more than a decade. However, Jacquemin was the first individual to get involved and gain visibility in the Chernobyl debate without being associated with a pro- or anti-nuclear group. His account focussed on the health effects of Chernobyl in France and claimed that the fallout was responsible for the increase in thyroid diseases. According to him, the health effects of the fallout were a taboo subject, and one which was continually avoided even after the French public authorities admitted that the country had indeed been hit by radioactive fallout.[193] He also asserted that this issue remained unaddressed, despite the fact that doctors in Corsica and cancer registers, according to him, clearly demonstrated dramatic increases in the incidence of thyroid diseases since Chernobyl.[194] Jacquemin's claims did not go unheard: a second edition of his book was released the following year.[195] In this edition, Jacquemin included a section on the AFMT, asserting in a self-congratulatory tone that the formation of this group was directly linked to his first book.[196]

Jacquemin's next book came out in 2001: *Tchernobyl: Aujourd'hui les Français malades* [197] (Chernobyl: today the French diseased). It opened with a clear statement: 'Fifteen years after the Chernobyl catastrophe, more and more French people suffer illnesses that are attributed to the radioactive cloud. The facts, the numbers are there. But the denial by the all-powerful nuclear lobby continues and persists'.[198] For this reason he pressed on with his endeavour: someone needed to give a voice to the sick people of France, whose existence had been denied and forgotten. To do so, he chose a pre-existing model: Alexievich's book *La supplication.* He presented the lives of individual 'French Chernobyl victims' in three or four-page testimonials over more than eighty pages. The title of this chapter was a statement in itself: 'Les témoignages des malades – La supplication des Français' (The testimonies of the diseased – the supplication of the French). But adapting the title of Alexievich's famous book charting the life stories of Chernobyl victims in Belarus and Ukraine was not the only element Jacquemin borrowed to project Alexievich's apocalyptic Chernobyl narrative onto France. He appropriated the entire criticism launched internationally by anti-nuclear activists against the official evaluation of the health situation in the most contaminated areas in Eastern Europe and applied it to France: suggesting that the same mechanisms had been employed in France to cover up the real impact of the accident. By applying the debate on Chernobyl's health effects in Eastern Europe to France, Jacquemin concluded that Chernobyl had not only increased thyroid diseases in France, but also various other cancers, such as lymph gland cancer, breast cancer, leukaemia and lung cancer.[199]

Jacquemin's mission did not end here. By 2002, his next book was already on the shelves: *Tchernobyl, conséquences en France. J'accuse…!*[200] (Chernobyl, consequences in France. I accuse…!). This time, in his choice of title, he did not just connect the Chernobyl health effects in France and in Eastern Europe, but also linked the French Chernobyl debate with one of the largest political scandals France had ever experienced, the *Dreyfus Affair*, presenting himself as a successor to Emile Zola. As in earlier works, Jacquemin presented the denial of French Chernobyl victims as the result of global nuclear policies and the 'sainte alliance'[201] (holy alliance) between the WHO and the IAEA.

Jacquemin's publications were of prime importance to the French Chernobyl debate. They not only fostered the judicial claims against the 1986 government and the radiation protection authorities, but also successfully discursively embedded the debate on Chernobyl's health effects in France within the wider international debate about the past and continuing health effects of the accident in Eastern Europe. In Jacquemin's narrative, the personal affectedness of France became its very own apocalypse.

Chernobyl Solidarity Movement

The term 'Chernobyl solidarity movement' refers to the collective non-governmental initiatives that provide humanitarian aid to the Eastern European regions most affected by the radioactive fallout. These groups, usually established through an individual person's initiative, are known to a wider public through organizing holidays abroad for 'Chernobyl children' and collecting clothes, medicine and presents for these children. Furthermore, many of these initiatives collect money that is invested in the infrastructure of hospitals and orphanages. The solidarity movement's activities have only recently become a topic of academic research.[202] In 2011, the first attempt at a more general account of the solidarity movement was undertaken by the German NGO Internationales Bildungs- und Begegnungswerk Dortmund (IBB, Association for International Education and Exchange). On behalf of the IBB, Isolde Baumgärtner compiled the publication *Tschernobyl und die europäische Solidaritätsbewegung*[203] (Chernobyl and the European solidarity movement). It included a history of the movement from a wider perspective and chapters on the various solidarity movement groups' activities in European countries. Most of these country reports were written by exponents. Linda Walker and Victor Mizzi, key actors in the British movement, contributed to the UK chapter.[204] However, since none of the French exponents contributed, I was asked to write the chapter on France.[205] For this purpose, Isolde Baumgärtner and the IBB gave me valuable information on French solidarity groups. In the same year it published this book, the IBB organized the International Partnership Conference in Minsk, aiming to connect the various initiatives arising all over Europe and to provide a forum for them to exchange ideas and knowledge.[206]

Writing the chapter for *Tschernobyl und die europäische Solidaritätsbewegung* was the first time I had taken a closer and more structural look at these groups' activities in France. I had only taken into account the group Enfants de Tchernobyl Bélarus (ETB) for my book *Tschernobyl und Frankreich*. However, I was more interested in the people behind this organization (Vassily Nesterenko, Galia Ackerman, Wladimir Tchertkoff, Solange Fernex and Michel Fernex) than in the solidarity movement as such. This is because the groups do not play a prominent role in the French public debate on Chernobyl. They exist and do their work, but do not shape the public image of Chernobyl. Furthermore, they are not big network organizations with intensive fundraising activities like in Britain. Moreover, there are also other local groups belonging to the Familles Rurales association that organize holidays for children from the worst affected areas.[207] Thus, the French solidarity movement groups are less visible than their British counterparts and are mostly only known in their local environment. One exception to the rule is ETB. However, this group is also an exception as far as solidarity movement groups are concerned. The ETB does not organize recreational stays for children in France. It mainly collects money for Belrad (Belarusian institute for radiation safety), an organization that was founded by Vassily Nesterenko, also a founding member of the ETB. Until his death in 2008, when his son Alexey took over, Nesterenko was the director of Belrad. Vassily Nesterenko established Belrad in 1990 as an independent organization to measure radioactivity levels in the people and foodstuffs within the areas of Belarus most affected by the fallout. Belrad has set up numerous local radiation measuring stations where people can have their own radioactivity levels checked as well as the radioactivity levels of the locally grown food. Belrad was thus able to collect a large amount of data that Nesterenko used for his own research into the health effects of radiation, particularly on children.[208] Belrad caught the attention of others outside Belarus mostly because of its use of pectin pills to reduce radioactivity in children, a 'cure' invented by Nesterenko. Belrad sells these pills under the trademark Vitapect in Belarus, but also internationally to host families of 'Chernobyl children' and, after Fukushima, to people in Japan. The ETB promoted the sale of these pectin pills.[209] Meanwhile the pills have become a highly political issue: while physicians strongly dispute the agency and the potential side-effects of the pectin pills, their price has become another controversial topic. Opponents claim that the pills are an effective way for Belrad to make money. The advocates of the pectin pills, on the other hand, consider the criticism the result of a concerted action on the part of the pro-nuclear lobby, continuing its attempts to block any research that might make life easier for radiation victims by denying their status as victims, as well as the existence of health risks stemming from the systematic ingestion of low doses of radionuclides.

Through its involvement in the pectin pills debate, the ETB has also become an important actor in the transnational debate on Chernobyl's health effects. At the same time, the ETB's activities have not been limited to Chernobyl: some of the ETB's founding members have played an important role in the general debate on the health effects of low-level radiation exposure. The ETB itself was a founding member of Independent WHO, the initiative that campaigns against the so-called IAEA–WHO agreement. The ETB's founding father and long-time president, Michel Fernex, is a driving force behind Independent WHO, and also responsible for initiating the permanent vigil at WHO headquarters in Geneva. Some years ago, Yves Lenoir, one of the most prominent French anti-nuclear activists and coauthor of *Tchernobyl-sur-Seine*, succeeded Michel Fernex as president of the ETB and Wladimir Tchertkoff and Alexey Nesterenko became vice-presidents. The prominence of these actors explains why the ETB is much more visible than other solidarity groups in the French Chernobyl debate.

The Chernobyl-related activities that the members of ETB carry out are directly linked to a clear anti-nuclear statement. This is, however, not true of all French Chernobyl solidarity movement groups. Other groups identify their work as humanitarian aid, and some even explicitly distance themselves from debates on nuclear politics. One example is the Fédération Echanges France Ukraine (FEFU). The FEFU is an umbrella organization uniting sixteen associations from all over France dedicated to cultural exchanges between France and Ukraine. Their humanitarian engagement is primarily aid for orphanages in Ukraine, but also organizing recreational stays in France for these orphans. The FEFU bases its mission on the general socio-economic situation of Ukraine, of which the 'Chernobyl children' are only one of many aspects; the focus is clearly cultural exchanges. For some of its member associations, however, helping 'Chernobyl children' is the heart of their activities.[210] Two other organizations working with 'Chernobyl children' are Accueil des Enfants de Tchernobyl and Les Enfants de Tchernobyl, both based in the Alsace region of France. These two associations are dedicated to improving the living conditions of children in the most contaminated areas, especially Ukraine. Ever since their foundation in the early 1990s, they have organized donations of medicine and clothes as well as sponsorships of individual children and recreational stays in France. By organizing events, they also raise awareness of the plight of the 'Chernobyl children'. Another organization, also called Les Enfants de Tchernobyl but based in Paris, has concentrated on providing medical aid in Ukraine. The association's founders, Dr Marie-Laurence Simonet and Dr Alexandra Moutet, initiated collaborations between French and Ukrainian hospitals. In addition, they fund the Centre Médical Français, based in Kiev paediatric hospital no 6.[211]

With the exception of the ETB, these solidarity movement groups have not much influenced the French Chernobyl debate; in fact, they have been barely visible. Only some of the associations have a website, and none of them has ever published a book or any other work beyond its close circle of members. This setting, that is to say the role and visibility of solidarity movement groups in the national Chernobyl debate, is completely different in Britain.

When the British actors in the nuclear sector and anti-nuclear side lost interest in Chernobyl in the early 1990s, another group of actors, the solidarity movement, picked up this topic and filled the 'discursive gap' in the years to come. At the two networking events which the IBB organized in 2011 and 2012 for the various European Chernobyl solidarity groups, I had the opportunity to talk to various people actively involved in the British movement. What I found astonishing about their accounts on fundraising activities, was the detachment of the Chernobyl topic from a more general debate on nuclear energy. I got the impression that in Britain, a commitment to helping 'Chernobyl children', through donating money or hosting holidays, does not automatically translate into an anti-nuclear stance. These activities can be rather equated with the Christian-oriented engagements to help underprivileged and handicapped children in less industrialized countries. Having researched publications written by people active in the British Chernobyl solidarity movement and the newspaper reports on the tenth and twentieth anniversaries of the accident which prominently featured their activities,[212] I was able to consolidate my initial hypothesis on the way 'Chernobyl children' are perceived in Britain, which is removed, separate and distinct from the nuclear discourse.

Noteworthy is the British solidarity movement's strong connection with Irish activists, particularly Adi Roche. Roche, a peace and environmental activist and candidate for the 1997 Irish presidential elections, founded Chernobyl Children International (CCI) in Cork back in 1991. Inspired by this initiative, Linda Walker launched Chernobyl Children's Project (CCP) UK in Manchester in 1995. Roche and her organization are special in that from the very beginning, she was highly successful in communicating her work to a wider audience. When Roche travelled to Chernobyl for the second time, she took along a whole film crew, transforming her experiences in the orphanages and the restricted zone around the nuclear plant into a film documentary. The result – *Black Wind, White Land* – was produced by Ali Hewson. Hewson, like Adi Roche, is an Irish peace and environmental activist and is married to U2's lead singer Bono. Hewson has been a strong supporter and lobbyist for CCI and arranged for all the profits from the U2 song 'The Sweetest Thing' to be donated to CCI.[213] Ali Hewson and Bono have also supported Adi Roche's work on other occasions. *Black Wind, White Land* was not the only medial output of Adi Roche's work; for her third visit to Chernobyl, Roche was accompanied by two Icelandic film crews. This resulted in strong Icelandic support for Roche's organization.[214] Furthermore, Roche used her experiences of travelling to Chernobyl and lobbying to bring 'Chernobyl children' to Ireland for a book: *Children of Chernobyl: The Human Cost of the World's Worst Nuclear Disaster*[215] was published in 1996 by a London-based publishing house. Although the book helped to spread news of her work, Adi Roche and her organization were already known in Britain. In 1994, Roche was contacted by Sarah Ferguson, the Duchess of York, who in the years to come supported CCI's work by providing funds through her foundation Children in Crisis and by initiating an international publicity and fundraising campaign for the 'children of Chernobyl'.[216]

Roche's publications and activities made her a central reference point in the British Chernobyl debate; and she also brought 'Chernobyl child' Igor to this debate. He had become the face – or rather the body – of Chernobyl by the mid-1990s. To the readers of her 1996 book, Adi Roche introduced Igor as follows:

> Igor Pavlovets was born on 3rd March 1987 and became the first acknowledged 'deformed' victim as a direct result of Chernobyl. His mother, in despair and shock at the sight of his broken body, abandoned him at birth to the state authorities. He was four years old when we first met him and had never seen anything beyond the four walls of the institution. Igor was special from the first moment I met him, not only because of his physical difficulties but also because his bright and loving personality overcame his limited circumstances. Igor, along with 60 other children, were being kept in this 'holding place' which had little medicine, food or special facilities.[217]

In order to prevent Igor's transfer to an adult mental asylum, Roche sought help from another 'Chernobyl children' aid group: the English organization Chernobyl Children's Lifeline (CCLL), founded by Victor Mizzi in 1992. The CCLL is the umbrella organization for 160 local groups spread throughout the UK. Between 1992 – when the first recreational stay was organized for twenty-two children from Belarus – and 2011, the CCLL has brought more than 46,000 children to the UK. To properly manage the 3,500 children coming every year, the CCLL runs an office in Minsk. In addition to these holidays, the CCLL collects money to support the families and improve the infrastructure of schools and hospitals in Belarus.[218]

Mizzi arranged for Igor to be brought to the UK, where he underwent a series of operations, was adopted by an English family and became a celebrity. He starred in two TV productions: one told the story of his move and the other, a couple of years later, his new life in England. His life was depicted as the ultimate success story: he had escaped the dark life of a miserable Belarusian orphanage to now enjoy the splendour of British medical care and the love of a whole nation. This narrative found its incarnation in Jane Warren's book *Igor: The Courage of Chernobyl's Child*,[219] first published in 1996. In this book, Jane Warren, an exponent of the CCLL, described in detail Igor's life in England and embedded this story in a general account on Chernobyl. Two aspects in particular merit closer examination: the interpretation of the (post-)Soviet state of development and how the narrative connected to broader nuclear issues. Warren framed the events of 1986 as derived from the political system and therefore as a Soviet accident. When describing the plant operators, she asserted that 'they were told repeatedly during their training, a nuclear power plant cannot explode'.[220] Her description of the problems adhered to a clear East–West divide: 'They [Soviet nuclear engineers] boasted that their nuclear reactors were of a superior design, despite the fact that British experts had condemned reactors like the one at Chernobyl as unsafe thirty years before the disaster'.[221] The way Warren described the work of the emergency crews and 'liquidators' underpinned her narrative of a poor Soviet understanding of the highly sensible matter they were dealing with: 'Everything was badly buried by men who didn't understand the need for absolute precision'.[222] Warren presented a similar lack of understanding regarding the evacuated people. Despite the fact that the evacuees included the inhabitants of Pripyat, people who were part of the USSR's scientific elite, her comments on the evacuations were all encompassing: 'These were country people. … They hadn't studied science. They didn't understand radiation'.[223] These comments reflect the implicit meta-narrative that coloured Warren's account: the image of simple, uneducated people betrayed by their political leaders. In this regard, Chernobyl was framed as a political rather than a nuclear accident. The impact of the nuclear fallout was considered devastating, and conveyed using apocalyptic language, for example to describe the orphanages: 'Gradually the dormitories were filling up with profoundly damaged babies and children. Many of them had defects caused by the radioactive food eaten by their mothers'.[224] However, the geographical impact of the accident, according to Warren's descriptions, was quite limited. A map indicating the 'approximate extent of radioactive contamination' only showed parts of Belarus, Ukraine and Russia.[225] No mention of the Chernobyl fallout in Britain in the entire book. Warren described Chernobyl's impact as a reality unfolding far from home. Britain, on the other hand, was presented as a place not facing any of the problems little Igor had to deal with in Belarus. Consequently, the chapter on his departure to Britain was entitled 'The Escape'.[226]

The CCLL promoted this 'escape' as a big event, setting the stage for Igor's career as a British humanitarian aid celebrity. Even on his flight to Britain, Igor was already accompanied by ITV and Sky channel TV crews. Born with only one fully developed arm, two short legs and deformed feet, the first thing that awaited Igor in his new home country was the correction of his body. Warren spoke of this in her chapter 'The Arm'. The media's exploitation of Igor's arrival in the UK figured prominently in this chapter: 'By now a television crew was filming Igor's quest for his new bionic arm. Victor Mizzi had always felt that there was a story to be told about the courage of Igor. Through an acquaintance, he was introduced to the production company Zenith North, which was very keen to make a documentary about Igor's early life and rescue to the West'.[227]

The subsequent chapters gave detailed descriptions of Igor's daily life in England, including anecdotes of the odd behaviour he brought from his previous home. 'Igor had settled into British life remarkably well, but there were still occasional signs of his early life in an institution. … One day she [Barbara, Igor's host mother] hurriedly stuffed some of his underwear into his drawer and the next time she looked, it had all been neatly folded into piles and smoothed flat'.[228] In his new life, Igor was often followed by TV cameras, a fact that Warren presented in a quite positive light: 'One morning in September 1994 it was time for Igor's first day at school. The television company, Zenith North, had arrived early to film this momentous day in his life'.[229] Igor had become the face and body that represented the destiny of the 'Chernobyl children'. His life was narrated as a story with a happy ending. Photographs in the book of Igor's perfectly formed smiling face underpinned this narrative. The visual representation of his smiling face stood for the British Igor of the present, while his severely deformed body represented Igor's Soviet past. Igor's physiognomy perfectly fit what the media demanded: his deformed body was a shocking image, but not too shocking to scare away spectators as it was balanced by his well-formed, normally proportioned and cute face. Igor's body offered the British media a way to put across the issue of Chernobyl without having to confront their audience with the situation in the worst affected regions of Eastern Europe. At the same time, Igor's life could be told as a British success story without needing to mention the topics of nuclear policies or radioactive sheep.

Applying Melanie Arndt's analytic categories – the 'externalization' and 'internalization' of fear in the motivation of solidarity movement exponents – to the case of Igor, it becomes clear that fear was not the main reference category in the public representation of his life story. The reference was hope, the hope of rescuing these poor 'Chernobyl children' from their awful destiny and the poor medical and living conditions they had to endure in these orphanages, institutions that could not provide what these children needed: medical care and love. To bring them to the UK, where an entire country awaited them with open arms, was the key to their future. These fairy tales were the perfect subject for TV productions. And Igor's life story became the ultimate fairy tale, thanks to the attention he received: 'The Duchess of York invited him and a group of visiting Belarusian children suffering from cancer to Princess Eugenie's fifth birthday'.[230] The mediatization of Igor's private life was also a way for the CCLL to raise funds for their work with other 'Chernobyl children'. 'Victor [Mizzi] continued to receive letters and money from people who were reading articles and watching television items about Igor's progress. Victor always replied to each letter, regardless of whether any money was enclosed in the envelopes. Many letters were from children who found it easy to identify with Igor's lively manner'.[231] That all this attention and excitement directed towards his person may have been difficult for Igor to handle – after all, he had just been moved from a place he had never left, to a completely new environment where he did not even understand the language – did not seem to have been cause for concern. Warren attributed Igor's wish to avoid all the attention to shyness: 'But Igor was growing reticent about his experience as a media celebrity. … Barbara picked up on his shyness, and stopped showing him the magazine and newspaper articles that were published about him'.[232]

It is not at all my intention to discredit the activities of the CCLL or the solidarity movement in general. These people dedicate an incredible amount of time, energy and money to improving the living conditions of children in the most affected areas, a commitment that deserves recognition. But I do believe the way the lives and bodies of these children are exposed as part of a fundraising strategy should be viewed in a critical light. Similarly, advertising campaigns of (half)naked, starving children for fundraising purposes have been widely criticized for a long time. In the case of Chernobyl, exposing children's bodies is equally contentious as the photos often focus on physical and mental handicaps.

Warren presented an apocalyptic narrative of these children's health: 'The health impacts of the Chernobyl disaster are difficult to measure, but the government estimates that there are currently over one million children deformed, damaged or diseased as a consequence'.[233] 'A 1993 health survey examined 500 Belarusian children and found only one to be completely healthy'.[234] She openly framed the public exhibition of Igor's life story as a means to raise awareness in British society of the accident's health impact in Eastern Europe: 'Unbeknown to him, he is bringing attention to the plight of the other children of Chernobyl left behind on contaminated land, and has become a symbol of their courage'.[235] And indeed, the media hype over Igor was successful, not only in terms of fundraising but also in stirring broader public interest in the 'Chernobyl children': 'The Carlton documentary about Igor's life, "Igor child of Chernobyl", was shown on Tuesday 6 June 1995. ... Barbara's and Victor Mizzi's phones began ringing with offers of help and money'.[236] 'The duty officer at Carlton also confirmed that rarely had a program provoked so many phone calls'.[237]

The public interest described above was, however, short-lived. It seemed to be more a result of the tabloid press-style exploitation of Igor's private life, which provoked curiosity and sensationalism, than a wider public interest in the health effects of a nuclear accident. This detachment of Igor's life story from the discourse on nuclear energy was also an inherent element of Warren's book. Although she narrated the consequences of Chernobyl as an apocalypse, she did not question the nuclear sector in general. She criticized the Soviet political system and 'man's carelessness'[238] but not nuclear technology or the 'nuclear state' as such. Moreover, Warren made no connection whatsoever to the British nuclear sector. In Igor's life story, Britain was the saviour, but a saviour worlds away from the events that had damaged Igor's body.

This detachment of the 'Chernobyl children' from the nuclear discourse allowed for British solidarity movement groups to address a wider audience in their fundraising activities. Chernobyl became a purely humanitarian issue, disconnected from any anti-nuclear connotations. Helping 'Chernobyl children' in Britain meant helping disadvantaged children, young victims of a Soviet accident. Even a pro-nuclearist could donate money to helping these children, who were a symbol of the devastating consequences when nuclear technology was placed in the wrong hands, rather than a symbol of nuclear power being at fault. It was better to bring these children to Britain for medical treatment, proving that Britain was far more technologically developed than these Eastern European countries; yet another rationale underpinning the argument that such an accident could never happen in the UK, and so there was no need to worry about the British nuclear enterprise. Rather than addressing issues of nuclear politics, the British solidarity movement exponents focussed on their joy and satisfaction from working with 'Chernobyl children'; they described the care of disabled children in particular as extremely rewarding.[239]

To enable thousands of children from Eastern Europe to spend holidays in Britain, the solidarity movement groups had to rely on the support of thousands of host families. British newspaper articles published during the accident's anniversaries presented portraits of these families and their host children. These portraits often included calls for new host families and for donations to the charities. Support did not just come in the form of private donations and host families, but also companies and enterprises participated. For instance, twenty-one coworkers of the Nationwide Electronic Service Network, on the company's tenth anniversary, went to the Belarusian orphanage Zhitkovitchi to renovate the shower and laundry block. Their work was photographed by Ian Beesly and a book entitled *Orphans of the Fallout* was published in 2001.[240] The profit from the book sales would pay for further renovation works the company intended to undertake at Zhitkovitchi in the years to come.

The activities of the British Chernobyl solidarity movement groups and the huge support they received should be seen within the context of Britain's very strong charity culture. In Britain, charity initiatives support the most diverse causes, from local playgrounds to cancer research. Fundraising initiatives also take on myriad forms, from sponsored runs to coffee mornings with home baking. Thus, when comparing the entire French solidarity movement to its British counterpart, we must bear in mind the situation in France, where such charity initiatives are far less common. A comparison between the two should therefore not be from a quantitative perspective but should rather investigate the role of each country's solidarity movement in its national Chernobyl debate.

In France, the solidarity movement has played only a marginal role in the Chernobyl debate. Although various French groups organize recreational stays for 'Chernobyl children', these groups did not become actors in the public debate, of course with the noteworthy and prominent exception of the ETB. French media reporting on the anniversaries never focussed on the life stories of 'Chernobyl children' or host families. It instead covered the *affaire Tchernobyl* and the debate on the accident's health effects in France. In Britain, however, from the mid-1990s onwards, the topic of 'Chernobyl children' dominated the public Chernobyl debate. For instance, while Busby was left to self-publishing *Wings of Death*, the exponents of the solidarity movement published their books through bigger publishing companies, and their activities were widely covered in the media during the anniversaries of the accident.

The work of British solidarity movement groups became known first and foremost through accounts documenting the personal life stories of 'Chernobyl children' that featured in books, newspaper articles and TV documentaries. This attention should be seen within the context of a British media culture, which, particularly in the tabloids, prefers to focus on individual life stories. As shown above, the topic of 'Chernobyl children' was never related to global nuclear politics or British nuclear policies. The opposite can be said in France. The ETB is a highly political group and its founding members rank amongst the most prominent actors in the national and transnational debate on the health effects of low-level radiation exposure.

When we consider the strong British focus on holidays for 'Chernobyl children' in relation to less perceived personal affectedness, some interesting findings emerge: according to the logic underlying the recreational stays, Chernobyl's health effects were confined to a distant location, and a few weeks' holiday in clean British air with healthy British food allowed the children to detoxify their systems from the radionuclides accumulated in their bodies. This logic at the same time implied that the situation in the most affected regions of Eastern Europe was rather manageable: a couple of weeks every year in Britain were all that was needed to give these children a healthy future. The fact that members of solidarity movement groups went themselves, also for longer stays, to the most affected areas, implies that they did not perceive the living conditions there to be a direct threat to people's lives. Thus, although apocalyptic imagery is used to depict the situation in orphanages, hospitals and health conditions of children in Belarus, Ukraine and Russia, an apocalyptic Chernobyl narrative regarding current health effects of the lasting low-level radiation in the environment is only partly implied in these accounts. For there to be any perceived personal affectedness in Britain, the explicit presence of such a narrative on the severe health effects of low-level radiation is essential. Since personal affectedness has played only a minor role in the British Chernobyl debate, the activities of the charity groups fit well in the dominant public image.

In France, however, inviting 'Chernobyl children' to the Hexagon for holidays somehow contradicts the logic underlying the strong perception of personal affectedness: Chernobyl is there, in France. Indeed, various initiatives organize recreational stays in France, but their activities never quite match the dominant public image. The health impact of the accident in France has become the focal point of the French Chernobyl debate. Therefore, if an observable health impact does indeed exist in France, as many believe, the situation in the worst affected areas of Eastern Europe must be considered a true apocalypse, like the one described by Alexievich and Bandazhevsky, authors whose work is fully part of the French but not the British Chernobyl debate. Interpreting Chernobyl as an apocalypse due to the current health effects of the lasting low-level radiation that has permeated the environment, challenges the underlying logic of recreational stays: what would be the long term positive effects for children brought to another country for a few weeks when they had to return to their lives in the apocalypse sooner or later?

Conclusion

After the early 1990s, neither the public authorities, the nuclear industry, the anti-nuclear activists nor other critical voices published much on Chernobyl in Britain. It was as if Chernobyl disappeared from the British public nuclear discourse. This phenomenon is probably linked to profound changes in British nuclear policies at that time. Within the framework of privatizing the British energy sector, the British government decided to stop building new nuclear power plants. The reactor at Sizewell B, where construction works had already begun in 1988, was completed and connected to the grid in 1995. But rather than being the pioneer heralding the arrival of a whole fleet of PWRs on British soil – as originally intended – the government decided in May 1995 that new build projects for nuclear power plants would no longer receive public (financial) support. Consequently, CEGB renounced its plans to build new nuclear plants. In the years to follow, the old Magnox plants were shut down – Oldbury and Wylfa were operating until the early 2010s – and so the share of electricity generated in Britain using nuclear energy steadily declined. The remaining nuclear power plants were eventually sold to EDF Energy, the British branch of the EDF Group. This change in Britain's nuclear policies may explain why British anti-nuclear activists lost interest in Chernobyl rather than instrumentalizing the topic in the fight against the civil nuclear enterprise, as was done in France. In the mid-1990s it looked like the British anti-nuclear power plant campaigners had got what they wanted: no nuclear expansion, and it would only be a matter of time until nuclear power plants would disappear from British soil.[241] However, the problem of Sellafield still had to be resolved as operations at the reprocessing plant were ongoing. Therefore, Sellafield continued to be the primary target of British anti-nuclear arguments. In France, however, the reprocessing plant in La Hague was not the only thing troubling anti-nuclear proponents. The trajectory of nuclear policies in France in the 1990s and early 2000s was essentially the opposite of what was happening on the other side of the Channel. The French government reconfirmed its all-nuclear policy and decided that the old PWRs should be replaced with a new generation of reactors, the EPR. But the government's nuclear commitment was not limited to France: the EDF and Areva were on track to become world leaders in nuclear power generation and engineering. This policy of strong political and financial support to the French nuclear power industry instigated the severe criticism of anti-nuclear activists and led to an intensification in protest and campaign activities, one of which was creating the Réseau sortir du nucléaire. Anti-nuclear campaigns were not just directed at the new build plans but manifested a protest against the entire industry, including the effects of uranium mining – in France and in Africa – or the working conditions of contract workers in French nuclear power plants. The memory of the *affaire Tchernobyl* continued to invoke strong criticism of the existing power dynamics within the French nuclear techno-political regime, and thus the memory of the Chernobyl accident conformed well to the general perspective of anti-nuclear protest. Therefore, French anti-nuclear campaigners shared a strong interest in preserving the memory of Chernobyl. In Britain, however, where the anti-nuclear discourse was dominated by the CND, it was not really possible to integrate the accident in an anti-nuclear weapons argument, which progressively focussed more on the threat of proliferation. Furthermore, the public focus on the concerns of the anti-weapons movement diminished considerably once the Cold War came to an end. Similarly, the number of people engaged in the British anti-nuclear energy discourse – including those interested in keeping the memory of Chernobyl alive from an anti-nuclear perspective – shrank. From the mid-1990s, this 'commemorative gap' in Britain was filled by the charitable organizations associated with the Chernobyl solidarity movement. However, these humanitarian groups have often refrained from taking a specific stance on nuclear energy, or else do not openly communicate their stance to garner support from the broadest possible group of potential donors in their fundraising campaigns. Therefore, in Britain, Chernobyl became associated solely with the destiny of a group of children suffering in Belarus, Ukraine and Russia. Moreover, often no further differentiation was made between Chernobyl and the Soviet system regarding which of them was responsible for the tragic situation.

In France, the lively debate about French Chernobyl health effects ensured that the accident remained a prominent topic in public discourse as well as in media reporting. The debate on personal affectedness in Britain, however, was marginal. People today in Britain might be rather surprised that there even was such a debate, however fleeting, about Chernobyl's impact on their health. This differs remarkably with the French case, where the fallout was increasingly considered to be a real and actual threat to public health. From the British point of view, Chernobyl had happened far away and stayed far away, only coming to the country in the form of the 'Chernobyl children'. The continued restrictions on sheep farms were soon forgotten, not least because they barely made it into news reporting. Thus, unlike Jean-Michel Jacquemin in France, a British activist like Christopher Busby could not base his argument and gain support from a common national Chernobyl memory and awareness.

What further strengthened the Chernobyl debate in France was the setting wherein perceived personal affectedness implied a certain shared destiny between the French people and the people living in the worst affected areas of Eastern Europe. This might explain how well the French received the work of Eastern European Chernobyl activists such as Alexievich, and their public support, as in the case of Bandazhevsky. Since no perception of personal affectedness exists in Britain, the Chernobyl narratives by Eastern European activists were not considered to be connected in any direct or actual way to British daily life; therefore, Eastern European activists did not enjoy the same level of success and recognition as they did in France. Consequently, these strongly apocalyptic narratives impacted the British Chernobyl debate to a far lesser degree.

The main reason why the debate on the health effects of the fallout in Britain remained marginal is probably because trust in the official experts and scientists essentially remained intact. Though they had been wrong in their predictions, better evaluations were eventually made available, demonstrating to the public that the science was making improvements.[242] The case of Benbecula illustrates this unbroken trust in science and public expertise: in fact, the moment scientists dismissed the claims made by local physicians, the topic completely disappeared from the newspapers. The same may be said of Busby and his work: the fact that his arguments were dismissed by scientists working within public authorities proved to considerably hinder their dispersion. But in France, the dynamics worked in the opposite direction. Here too, public authorities dismissed Jacquemin's claims, but their rejection seemed to only make his claims appear all the more valid; the rejection of his work is what turned it into proof that the nuclear experts were continually covering up the 'truth' about Chernobyl. This perception is still valid today: in the panel study on risk perception which the IRSN conducts every year, Chernobyl continuously holds the lowest credibility ranking of official evaluations on risk situations. In 2017, only 8 per cent of the people interviewed for this representative panel study believed that the truth is being told about Chernobyl's health effects in France.[243]

In order to fully grasp why the British Chernobyl debate dried up in the early 1990s, we must also consider the settings and contexts that lie outside the immediate Chernobyl debate. But because the history of the British Chernobyl debate from the mid-1990s to the early-2000s is mostly a history of a non-debate, it is far more difficult to connect the Chernobyl debate to wider societal issues than in France, where these connections are openly expressed in various actors' arguments. Despite this lack of sources, we can identify some particularities of British (political) culture that may have caused the almost total disappearance of Chernobyl from the public discourse. One such particularity is the British media system. The British newspaper market is dominated by the tabloids, and thus the currency of news value is of higher importance than in other markets. This logic might explain why, for instance, newspaper reporting on Benbecula did not probe further into the wider debate on Chernobyl-related health effects and only covered this topic as long it provided a catchy headline. Furthermore, the British style of the broadsheet's news reporting insists on a clear distinction between facts and opinion. Thus, it is far less common in the UK than in France for a journalist to openly take a political stance on the topic they are discussing. For this reason, the French media system made it easier for journalists like Hélène Crié, Noël Mamère or Galia Ackerman to play an active role in the Chernobyl debate and use their position to convey their own opinions and narratives on Chernobyl to a wider public. Naturally, there is always an exception to the rule. Comparing French and British journalists' engagement in the Chernobyl debate, this exception is Rob Edwards.[244] Edwards, who in the 1970s and 1980s campaigned for SCRAM and the CND, has managed to be an activist and a journalist at the same time. Furthermore, similarly to Hélène Crié, he wrote several books openly expressing his anti-nuclear arguments. In the British setting, Edwards is a rather singular case, not only because he is one of the very few journalists who had continued to write about Chernobyl,[245] in the early 2010s, he was the only prominent British journalist who actually wrote on environmental issues from an explicitly anti-nuclear standpoint.[246]

Another particularity of British (political) culture that may have hindered the expansion of the Chernobyl debate is the British political party system. Like its French counterpart, the UK Green Party was the only political party to entirely oppose nuclear power.[247] The difference is the role each party played in their respective country's political system. Compared to other political parties, the Greens are not a major force in France or in Britain and are primarily successful in local or European Parliament elections. But whereas Les Verts (the French Greens) were included in the 'Plural Left' government in 1997, the 'British Greens'[248] only won their first seat in the national Parliament in 2010, some thirteen years later.[249] Moreover, some of the key actors in the French Chernobyl debate are successful Green politicians, for example Noël Mamère and Michèle Rivasi; and the *affaire Tchernobyl* forms an essential part of their political identity. Thus, in France, the Chernobyl debate has continued to have a political representation in the form of Les Verts. Regarding the *affaire Tchernobyl*, however, Les Verts opposed the entire French political sphere: after all, the 1986 government, which was accused of covering up the true impact of Chernobyl, was the *Première cohabitation* with François Mitterrand as

President of the French Republic and his political opponent Jacques Chirac as Prime Minister. In Britain, however, the divide created over the issue of proper emergency management in the spring of 1986, ran across party lines, not only cutting through the Labour Party but also partly placed Parliament in opposition to the Government. Although this dispute was settled with the House of Commons Agriculture Select Committee report on the Government's reaction in 1988, some Labour MPs like David Clark continued to broach the issue that sheep from the affected areas had slipped through and been taken to market. It would be an oversimplification to say that the British political party system as such had hindered the expansion of the Chernobyl debate. Although the weakness of the Green Party may have been a hindering factor, the power relations between the British Parliament and the Government, and the political opposition between the Labour Party and the Conservative Party in 1986 could, in theory, have been the arena for expanding the debate.

When I discussed my research, many people pointed to the general depoliticization of the British public generated by Thatcherism and the neoliberal turn of the Labour Party. In their opinion – the people who raised this issue were mainly anti-nuclear minded – this depoliticization had led to a decline in the British anti-nuclear movement and transformed nuclear power issues into non-issues, effectively removing them from the sphere of public debate. This study is not equipped to verify this hypothesis. To explain the lack of sources with a general non-interest in politics or in particular, a non-interest in nuclear politics, albeit tempting, does not constitute a scientific argument. This does not necessarily mean this hypothesis is not valid, only that I cannot reach or corroborate this conclusion from the material I researched. However, ignoring this argument would exclude an essential part of the contemporary British anti-nuclear discourse from my analysis, which is why I mention it here.

When we compare how the French and the British deal with and debate Chernobyl, it is interesting to note that the impact of the accident has become a social-philosophical issue in France – primarily in the work of the Caen-Group – whereas in Britain, there is a more hands-on approach to the topic. This applies to the national level relating to how the government managed the issue of contaminated sheep as well as the transnational level in terms of the practical help for 'Chernobyl children'. However, any attempt to explain this phenomenon runs the risk of drawing heavily on national stereotypes. Although it might be tempting to refer to the French love of existentialism to explain why the narrative of Chernobyl as 'the end of all common reference points' (Lemarchand/Grandazzi) has been so widely received in France, this would barely scratch the surface of the Chernobyl debate. To explain the different trajectories of the two national debates through national cultural mentalities would run the risk of getting trapped in the pitfall of using highly biased and constructed stereotypes of Britishness and Frenchness. Emphasizing such stereotypes would imply that less attention was paid to the power structures at stake in the Chernobyl discourse and to the power structures specific to the national nuclear discourses where the Chernobyl discourse is embedded, and thus be counterproductive to my research.[250]

To explain the quasi-non-existence of the British Chernobyl debate, we should not just consider structures: the comparison with the French case highlights the equal importance of individual actors. Had Alexievich's narrative not been promoted by people like Galia Ackerman, Frédérick Lemarchand and Guillaume Grandazzi, it would never have become so popular in France. The same is true for Bandazhevsky's work: had the CRIIRAD coworkers and especially Michèle Rivasi not intensively disseminated information about Bandazhevsky and his theories on the health impact of exposure to low-level radiation, his work and imprisonment would not have received the same amount of attention. Similarly, with the topic of personal affectedness: without Jean-Michel Jacquemin's publications, fewer people in France may have related their thyroid diseases to the Chernobyl fallout. Perhaps Christopher Busby could have had a similar 'career' in Britain. However, for him, the impact of Chernobyl in Britain was only ever a secondary issue and he instead became known for his views on health effects from the emissions of the nuclear installations in Britain and weapons testing fallout. This handful of examples reveals the importance of individual agency in the process of politicizing a certain topic. Discursive power relations and wider political and cultural structures are only one side of the story, the other is whether there are actors willing to take up the topic; or if potential actors focus their time and energy on other issues. In Britain, an immense amount of time and energy has been invested in the fight against nuclear arms: the Aldermaston Women's Peace Camp, which has existed since 1985, might be the most prominent example of this ongoing commitment. It is striking that in France, by comparison, there is at best only a marginal debate on nuclear weapons: the French secret service's bombing of the Rainbow Warrior in 1985 had just as negligible an effect on François Mitterrand's career as Jacques Chirac's commitment in the mid-1990s to French nuclear weapons testing in the South Pacific had on his re-election.

Notes

1. These dissident voices have not yet been subject to a broader scientific research project. As a first move in this direction, Thomas Bohn provides background information on the main authors: Bohn, 'From Recording the Catastrophe to Tackling the Trauma'.
2. Medvedev, *Nuclear Disaster in the Urals.*
3. A French translation of his book appeared much later, in 1988: Medvedev, *Désastre nucléaire en Oural.*
4. Medvedev, *Legacy of Chernobyl.*
5. Ibid., 165.

6. Ibid., 166.
7. Ibid., 130.
8. Ibid.
9. Ibid., 166.
10. Robert Peter Gale's book is another example. Gale, an American doctor, carried out bone marrow transplants on Chernobyl firefighters in Moscow during the weeks after the accident. This work was financed by the American millionaire Armand Hammer. In 1988, Gale published *Chernobyl: The Final Warning*, combining his diary from his time in Moscow with accounts on the probabilities of nuclear accidents occurring and the problems with large-scale evacuations. Gale's surgery work in Moscow features in most Chernobyl accounts written in English. It does not feature, however, in French Chernobyl narratives nor does the Bibliothèque nationale de France have a copy of Gale's book.
11. Medvedev, *La Vérité sur Tchernobyl.*
12. Medvedev, *The Truth about Chernobyl.*
13. Ibid., 57.
14. Ibid., 265.
15. Medvedev, *No Breathing Room.*
16. Ibid., 28ff.
17. Marples' first book on Chernobyl was published in 1987 already: *Chernobyl and Nuclear Power in the USSR*; his next work only a year later: *The Social Impact of the Chernobyl Disaster*. Through studying the impact of Chernobyl, Marples became increasingly interested in Belarus, which led to his third book on Chernobyl: *Belarus: From Soviet Rule to Nuclear Catastrophe.*
18. Yaroshinskaya, *Chernobyl: The Forbidden Truth.*
19. Yarochinskaya, *Tchernobyl, vérité interdite*. In 2004, the editors of *Les silences de Tchernobyl* included an article by Yaroshinskaya.
20. Yaroshinskaya, *Chernobyl*, 123.
21. Ibid., 2.
22. Ibid., 3ff.
23. Chernousenko, *Chernobyl: Insight from the Inside.*
24. Ibid., ix.
25. Ibid., 39.
26. Ibid., 171.
27. Ibid., 217.
28. Alexievich, *Voices from Chernobyl.*
29. For more on Alexievich's work, see: Kalmbach, 'Chernobyl as a National and Transnational Site of Memory', 142ff.; ibid., *Tschernobyl und Frankreich*, 123ff.
30. On 13 March 1996, a man killed sixteen children and a teacher at Dunblane primary school in Scotland, before shooting himself. This school massacre was intensively reported in the media at the time.
31. Frewer et al., *Media Reporting*, 17. The UK newspapers in the research are: *The Guardian, The Observer, The Times, The Sunday Times, The Sun, News of the World, Daily Mirror, Sunday Mirror, Reading Evening Post* and *Whitehaven News*, listed on: ibid., 25.
32. Frewer et al., *Media Reporting*, 38.
33. *The Daily Record*, 'Nuke Link to Cancer Ruled Out'.

34. Hunt, 'Chernobyl Link to Cancer Cluster'.
35. As far as I can tell from my research, the case of Benbecula was the first time the Chernobyl fallout in Britain was publicly claimed to have observable health effects.
36. *The Independent*, 'Experts to Investigate Island Cancer Deaths'.
37. *The Daily Record*, 'Fags to Blame for Isle Cancer Scare'.
38. Johnstone, 'Health Director to Study Big Rise in Benbecula Cancer'.
39. Buie, 'Cancer Island Inquiry Health Board'; MacDonald, 'Isle of Desolate Beauty'; Tinning, 'Chernobyl Fall-Out Denial'.
40. This quote by Dr Robert Kendell is in all the articles cited above published in the wake of *The Independent*'s article of 31 March 1996.
41. Western Isles Health Board officials checked all the medical records of general practitioners on Benbecula to determine if the incidence of cancer was statistically different from the average in the rest of Scotland.
42. The crisis management by Dr Robert Kendell is what Harry Collins and Trevor Pinch call a strategy of 'paternalistische Beschwichtigung' (paternalistic appeasement) typical of how British public authorities react to public health threats. Collins and Pinch consider that the British government's reactions to Chernobyl in 1986 and the BSE crisis exemplified this strategy. See: Collins and Pinch, *Der Golem der Technologie*, 168.
43. Tinning, 'Chernobyl Fall-Out Denial'.
44. Smith and Clark, *The Transport and Deposition of Airborne Debris.*
45. Ibid., 14.
46. Ibid., 37.
47. Ibid., 48.
48. Ibid., 49.
49. Ibid., 52.
50. Ibid.
51. Ibid., 2.
52. Ibid., 53.
53. Sandalls, *Hot Particles from Chernobyl*, 1.
54. See: Lhomme, *L'insécurité nucléaire.*
55. Department of Agriculture and Fisheries for Scotland, *Chernobyl Accident.*
56. Ministry of Agriculture, Fisheries and Food, *Radionuclide Levels*
57. Kane, *Chernobyl: Evaluation and Radiological Assessment.*
58. Cawse and Baker, *Survey of Radioactive Caesium.*
59. East, *Measurement of Radioactivity.*
60. Camplin and Great Britain Directorate of Fisheries Research, *Radioactivity in Freshwater Systems.*
61. Sandalls, *Hot Particles from Chernobyl.*
62. Bonnett, *Transport Mechanisms.*
63. Instead of framing the way the fallout was handled as 'neutral' or 'non-neutral', it makes sense to apply Gabrielle Hecht's concept of *nuclearity*, see: Hecht, *Nuclear Ontologies*. For applying this concept to Chernobyl, see the conclusion in: Bauer and Kalmbach and Kasperski, 'From Pripyat to Paris'.
64. Haywood and Institute of Physical Sciences in Medicine, *Response of Medical Physics Departments.*

65. Ibid., 26.
66. Nisbet, 'Management of Chernobyl-Restricted Areas'.
67. Worley and Lewins, *The Chernobyl Accident and Its Implications*.
68. Nisbet and Woodman, *Options for the Management of Chernobyl-Restricted Areas*. A summary was published a year later in the *Radiological Protection Bulletin*: Nisbet, 'Management of Chernobyl-Restricted Areas'.
69. Nisbet and Woodman, *Options for the Management of Chernobyl-Restricted Areas*, v.
70. Ibid., vi.
71. Ibid.
72. Soil measurements and aerial surveys establish the contamination of an area with a specific radionuclide (regarding the long-term contamination caused by Chernobyl: radiocaesium).
73. Cook, 'Render unto Caesium 134'.
74. In discussing Brian Wynne's work, Harry Collins and Trevor Pinch stressed that sheep farmers in Cumbria saw the scientists who evaluated the radioactive contamination as 'victims of the political pressure from the government' – a government waiting for this 'perfect pretext to shift the blame for the covered-up Sellafield contamination on to Chernobyl'; Collins and Pinch, *Der Golem der Technologie*, 167 (translated by the author).
75. Cook, 'Render unto Caesium 134'.
76. These articles were accessed through the databases ProQuest and Newsbank in the British Library, using the search key 'Chernobyl sheep farm'.
77. See for example: Snape, 'Chernobyl Pay-Out'; Ghazi, 'Chernobyl Fall-Out'; McKie, 'Britain's Deadly Chernobyl Legacy'; Ward, 'Hill Farmers Living under Cloud'; Wainwright, 'Ministers Deny Risk'; Clouston, 'Legacy of Night when it Rained Radiation'; Heath, 'Nuclear Cloud Hangs over the Hills'; Meikle, 'Chernobyl Legacy'.
78. IPSN, *Tchernobyl, 10 ans après*.
79. For an analysis of this brochure, see: Kalmbach, *Tschernobyl und Frankreich*, 93ff.
80. The following account is a summary of: Kalmbach, *Tschernobyl und Frankreich*, 95–9.
81. IPSN, *Conséquences radioécologiques et dosimétriques*.
82. Renaud et al., *Les retombées en France*.
83. Ibid., 140.
84. Ibid., iv.
85. IPSN and InVS, *Evaluation des conséquences sanitaires*. A summary was released at the same time: IPSN and InVS, *Evaluation des conséquences sanitaires – synthèse*.
86. Ibid., 17.
87. Some anti-nuclear activists blame this increase on the fallout from global weapons testing.
88. For a detailed analysis of this court case, see: Kalmbach, *Tschernobyl und Frankreich*, 116ff. The court's decision in September 2011 to dismiss the claim evoked strong public criticism, which was widely covered by the French media (see for example *Le Monde* and *Libération* on 7 September 2011). Not only was the decision contested but also how the case was conducted and the dismissal of *juge d'instruction* Marie-Odile Bertella-Geffroy, who had taken the plaintiffs' arguments seriously and invited experts outwith the French nuclear sector to testify. After ten years of investigations she was dismissed in March 2011 and, within months, the case was dismissed.

89. Original wording: 'Défaut de protection des populations françaises en général et des groupes à risques en particulier contre les retombées radioactives de l'accident de Tchernobyl'.
90. CRIIRAD and AFMT, *Communiqué joint.*
91. Since the late 1990s, it has become more common in France to bring claims against politicians and hold them legally responsible for the negative outcomes of their decisions, see: Roussel, 'Scandales et redéfinitions de la responsabilité politique'.
92. For the EDF's communication on nuclear power and in particular the campaign 'Aujoud'hui, 75% de l'électricité est nucléaire', see: Topçu, *L'agir contestataire*, 310ff.
93. The analysis of this brochure is adapted from: Kalmbach, *Tschernobyl und Frankreich*, 89–90.
94. EDF, *Tchernobyl: le vrai, le faux et l'incertain*, 2.
95. Ibid., 7.
96. Ibid., 6.
97. Ibid., 7.
98. Ibid., 10.
99. Ibid., 11.
100. Watt Committee on Energy, *Five Years after Chernobyl.*
101. For a comprehensive overview of the sources used for this report, see the reference list: ibid., 45.
102. Watt Committee on Energy, *Five Years after Chernobyl*, 26.
103. Ibid., 29.
104. Ibid., 32.
105. Ibid., 31.
106. Ibid., 33.
107. Ibid., 36.
108. Ibid., 33.
109. Ibid., 37.
110. Ibid., 44. The report gave clear guidelines at the beginning, p. viii on how to address the health effects of Chernobyl: 'Whether there has been a "cover up" is more difficult to determine; ultimately there has to be an assessment of the care that can be given to those who might be affected. It is not feasible to extend limitless funds on correcting some perceived defect and it is naïve to suppose that such an absolute can be sustained'.
111. Ibid., 17.
112. Belbéoch, 'Le complot international'. Translation by the author of a quotation from: Castanier, 'Contamination des sols français', 7.
113. Belbéoch and Belbéoch, 'Tchernobyl, une catastrophe'.
114. Belbéoch and Belbéoch, *Tchernobyl, une catastrophe: Quelques éléments pour un bilan* (Éd. Allia: Paris, 1993). The text is available online: http://www.dissident-media.org/infonucleaire/Tchernobyl_une_catastrophe_1993.pdf (last accessed 20 February 2020).
115. Belbéoch and Belbéoch, *Tchernobyl, une catastrophe*, 17 (online edition).
116. Ibid.
117. Ibid.
118. Medvedev, *La Vérité sur Tchernobyl.*

119. Yarochinskaya, *Tchernobyl.*
120. Rivasi and Crié, *Ce nucléaire qu'on nous cache*, 113.
121. CRIIRAD and Paris, *Contaminations radioactives.* For a detailed analysis of this publication, see: Kalmbach, *Tschernobyl und Frankreich*, 100–3.
122. CRIIRAD and Paris, *Contaminations radioactives*, back-cover of the book.
123. For more information on the AFMT and the manifesto directed against its activities, see: Kalmbach, *Tschernobyl und Frankreich*, 110–13.
124. *Libération*, 'Message aux malades de la thyroïde'.
125. Ibid.
126. Ibid.
127. Grandazzi and Lemarchand, *Les silences de Tchernobyl.*
128. For a detailed account of the 2004 edition of *Les silences de Tchernobyl* and more on the editors' background, see: Kalmbach, *Tschernobyl und Frankreich*, 125–9.
129. Grandazzi and Lemarchand, 'Avant-propos', 7.
130. Ibid., 10.
131. For further information on Yuri Bandazehsky, see chapter 3.
132. Grandazzi, 'L'atome en héritage', 129.
133. Alexievitch, *La supplication.* Galia Ackerman is a journalist and translator. She worked for many years for France Inter where she reported on Russia and post-Soviet countries.
134. Areva was established in 2001 when the companies Framatome ANP, Cogema and CEA-Industrie merged. This reorganization of the French nuclear sector was a government initiative to make the industry more competitive. Areva, as holding company, brought together the companies Areva NP, Areva NC, Areva T&D and Areva TA. Areva NP (NP = Nuclear Power) became a world leader in reactor design, development and construction, operating under the name Framatome ANP (ANP = Advanced Nuclear Power) until 2006. Starting in 2017, and due to intense financial difficulties, Areva was reorganized. Areva NP was outsourced, renamed Framatome and most of it taken over by EDF.
135. For an account of British environmental activism in the 1990s, see: Rootes, 'Britain'.
136. I was only able to read about this item in other sources. It is missing from the British Library. My request to FoE's London head office to consult the report in their archive or to put me in touch with an author was refused. The only information I received was that Patrick Green left FoE in 1996.
137. *The Guardian*, 'Chernobyl Bans Criticised'. Another reference to the FoE report is: *The Guardian*, 'Disaster That Fell with the Rain'.
138. Friends of the Earth, *The Chernobyl Legacy*, 5.
139. Ibid.
140. Cook, 'Render unto Caesium 134'.
141. *The Guardian*, 'Disaster That Fell with the Rain'.
142. Ibid.
143. Meikle, 'Chernobyl Legacy'.
144. Ward, 'Hill Farmers Living under Cloud'.
145. Park, *Chernobyl: The Long Shadow.* He wrote the book while he was a Professor of Geography at the University of Lancaster.

146. Gould, *Fire in the Rain*. At the time he was a Professor of Geography at Penn State University.
147. Rickard, *The Chernobyl Catastrophe*; Condon, *Chernobyl and Other Nuclear Accidents*; Paul Dowswell, *The Chernobyl Disaster*; Ingram, *The Chernobyl Nuclear Disaster*; Parker, *Chernobyl 1986*. These children's non-fiction books are an illustrated mix of action story, introduction to nuclear engineering and pedagogic sensitization to the risks of nuclear technology. Although they differ in what they present as 'facts', the authors all aimed to provide a balanced account informed by different sources, but mostly produced very superficial and contradictory narratives. For an analysis, see: Kalmbach, 'The Contested Truth', 182–9.
148. Read, *Ablaze*.
149. Rob Edwards also continued writing about Chernobyl, but just in newspapers. The only book he published after *Britain's Nuclear Nightmare* was on the health effects of Sellafield: Edwards and D'Arcy, *Still Fighting for Gemma*.
150. Mould, *Chernobyl Record*.
151. Ibid., xiv.
152. Ibid., xvii.
153. See: Topçu, *L'agir contestataire*, 305ff.
154. Mould, *Chernobyl Record*, 225.
155. The International Chernobyl Project, *An Overview: Assessment of Radiological Consequences*, 32.
156. Ibid., 33.
157. Ibid., 35.
158. Mould, *Chernobyl Record*, 231.
159. On 28 May 1959, the World Health Assembly approved an agreement between the WHO and the IAEA to regulate interaction. Alongside practical aspects such as co-operation in technical or administrative fields (Article VI), reciprocal representation (Article II) or collection and exchange of statistical data (Article VII), this agreement regularized co-operation and consultation between the two agencies regarding their subject-matter in general (Article I).
160. See: Independent WHO's website.
161. The fact that the Wikipedia article 'Independent WHO' exists only in French reflects this situation.
162. For a detailed account, see: Kalmbach, 'The Contested Truth', 279–87.
163. Mould, *Chernobyl Record*, 235.
164. Ibid., 260.
165. Ibid., 286.
166. The worst example of this Anti-Busby campaign was the website 'Chris Busby Exposed' that used personal defamation tactics to strategically damage his reputation; the website has since been removed. It was formerly accessible on: http://junkscience watch.wordpress.com/ (last accessed 15 June 2013).
167. George Monbiot is a British journalist who has worked for the BBC for years and is known not just for his investigative documentaries on health and environmental issues but also for his books on climate change. In the early 2010s, climate change became his main field of activism and he became a declared campaigner and supporter of nuclear

power. See the open letter Monbiot wrote with Stephen Tindale, Fred Pearce, Michael Hanlon and Mark Lynas to Prime Minister David Cameron in March 2012: Monbiot et al., 'A Letter to David Cameron'. Monbiot strongly attacked the anti-nuclear activists, claiming they were deliberately promulgating incorrect information in order to incite fear of radioactivity in people; see also his post-Fukushima blogpost 'Evidence Meltdown'.

168. Monbiot, 'Christopher Busby's Wild Claims'.
169. Wikipedia, 'Talk: Christopher Busby'.
170. Busby, *The Wings of Death*.
171. Ibid., 188.
172. Ibid., 91.
173. Ibid., 92.
174. Ibid., 137.
175. Ibid., 297ff.
176. Ibid., 298.
177. Ibid., 301.
178. Low Level Radiation Campaign, *Do We Really Want Nine Chernobyl Accidents Every Year?*
179. Ibid., 7.
180. Busby and Scott Cato, 'Increases in Leukemia in Infants in Wales and Scotland'.
181. Ibid., abstract.
182. Ibid., discussion.
183. NRPB, 'Response to Wind Blown Particles and Cancer Mortality'.
184. Ibid., 13.
185. This response was formerly available on the Low Level Radiation Campaign website: http://www.llrc.org/wobblyscience/subtopic/christopherrobinissayinghisprayers.htm (last accessed 15 November 2013).
186. Busby, *The Wings of Death*, 302.
187. Charpak and Garwin and Journé, *De Tchernobyl en tchernobyls*.
188. For an analysis of this book, see: Kalmbach, *Tschernobyl und Frankreich*, 140–44.
189. Ibid., 104–10. The following account summarizes the content of this chapter.
190. Jacquemin concluded all his books with the acknowledgement: 'Merci à DIEU et à mon Guide, Mélahel, pour cette nouvelle mission, pour Votre aide et Vos protections'. (My thanks to GOD and my Guide, Mélahel, for this new mission, Your aid and Your protections.)
191. See: Castanier, 'Contamination des sols français', 49.
192. Jacquemin, *Ce fameux nuage*.
193. Ibid., 163.
194. Ibid., 164ff.
195. Jacquemin, *Ce fameux nuage* (1999).
196. Ibid., 319ff.
197. Jacquemin-Raffestin, *Tchernobyl: Aujourd'hui les Français malades*.
198. Ibid., 17.
199. Ibid., 147ff.
200. Jacquemin-Raffestin, *Tchernobyl, conséquences en France*.
201. Ibid., 186.

202. The main literature includes: the works of Astrid Sahm on the impact of solidarity movement activities in Belarus (Sahm, 'Auf dem Weg in eine transnationale Gesellschaft'); Melanie Arndt on the motivations of West German solidarity movement groups (Arndt, 'Verunsicherung vor und nach der Katastrophe'); and Isolde Baumgärtner on forms of aid and projects across Europe (see chapters by Isolde Baumgärtner in: Internationales Bildungs- und Begegnungswerk Dortmund, *Tschernobyl und die europäische Solidaritätsbewegung*).
203. Internationales Bildungs- und Begegnungswerk Dortmund, *Tschernobyl und die europäische Solidaritätsbewegung*.
204. Walker, 'Die Arbeit von "Chernobyl Children's Project (UK)"'; Mizzi, 'Über die Arbeit von Chernobyl Children's Lifeline'.
205. Kalmbach, 'Die Wolke, die an der Grenze haltmachte'.
206. I was invited to take part in this conference which took place from 17 to 20 April 2011, on the twenty-fifth anniversary of Chernobyl, in the Internationale Bildungs- und Begegnungsstätte 'Johannes Rau' Minsk, an IBB branch in Belarus directed by Astrid Sahm at the time.
207. I am grateful to Marie-Hélène Mandrillon for this information.
208. For further information on Vassily Nesterenko and Alexey Nesterenko, see: chapter 3.
209. An ETB brochure describing the organization's activities, said the following about pectin pills: 'It is possible to help the sick children by financing apple pectin cures. Apple pectin is a natural adsorbent and the pills can eliminate the caesium 137 more rapidly from their bodies. The radionuclide caesium 137 is present in their daily nutrition (50% or even 70% of the total amount of accumulated radioactivity is eliminated within 3 weeks of the cure when repeated every trimester)'. In: Enfants de Tchernobyl Bélarus, 'La plaquette réactualisée de l'association'.
210. For further information on the FEFU and its member associations, see the 'Portrait' on FEFU's website.
211. This paragraph is adapted from Kalmbach, 'Die Wolke, die an der Grenze haltmachte', 85–6. I am grateful to Isolde Baumgärtner for information on these associations.
212. A search in the database Newsbank for the time span 15 to 30 April 1996 (tenth anniversary), delivered numerous articles on 'Chernobyl children' in local and national newspapers. These included portraits of individual children, accounts of certain solidarity groups or members and stories by host families. However, there was a distinct lack of articles on other Chernobyl-related topics, such as sheep farm restrictions.
213. Wikipedia, 'Ali Hewson'.
214. Adi Roche and the CCI's work was the theme of the film documentary *Chernobyl Heart*, which won the Academy Award for Best Documentary (short subject) in 2003.
215. Roche, *Children of Chernobyl*.
216. For an account of this cooperation, see: ibid., 107ff.
217. Ibid., 19.
218. These numbers are taken from: Mizzi, 'Über die Arbeit von Chernobyl Children's Lifeline'.
219. Warren, *Igor*.
220. Ibid., 9.
221. Ibid.
222. Ibid., 13.

223. Ibid., 18.
224. Ibid., 32.
225. Ibid., 14.
226. Ibid., 56.
227. Ibid., 68ff.
228. Ibid., 71.
229. Ibid., 78.
230. Ibid., 91.
231. Ibid., 92.
232. Ibid., 93.
233. Ibid., 105.
234. Ibid., 106.
235. Ibid., 107.
236. Ibid.
237. Ibid., 108.
238. Ibid., book cover.
239. In his 2003 book, also Irish Brother Liam O'Meara expressed the Christian ideal to care for the poorest as major incentive for helping the 'Chernobyl children'. O'Meara's account of working with these children reads like a therapy session. He described the happiness he derived from the grateful eyes of the children and the pain Irish people felt when they had to give back 'their' children at the end of their holidays. O'Meara detailed the children's suffering in Eastern European orphanages but scarcely mentioned the underlying nuclear accident: O'Meara, *Fallout*.
240. Beesley, *Orphans of the Fallout*.
241. This was, however, not the case. In 2006, the British government reversed the decisions taken in the mid-1990s and announced the construction of several new plants. The next chapter will deal with this development in more detail.
242. A statement on scientists' failure to predict the impact of Chernobyl on British sheep farming by Dr Brenda Howard of the Institute of Terrestrial Ecology illustrates this thinking: '"They [the scientists] couldn't really predict what would happen, which is determined by grazing pressure, type of vegetation and innumerable ecological and environmental factors," she says. "As it turns out, their predictions were wrong. It's as simple as that."' In: Cook, 'Render unto Caesium 134'.
243. IRSN, *Le Baromètre IRSN sur la perception des risques*, 36. See for an analysis of this panel study: Kalmbach, 'Point de vue'.
244. See his article: Edwards, 'Chernobyl Fall-Out'. He reported on an aerial monitoring survey, commissioned by *Scottish Eye*, a Channel 4 TV programme, which registered higher rates of radioactivity than on the official maps. He stated: these 'results are bound to rekindle the argument over the effectiveness of the Government's response to Chernobyl and its preparedness for any future accident'. Despite pointing to the 'allegation repeatedly made by Labour's agriculture spokesman David Clark, MP, that 100,000 sheep from high-risk areas went for public consumption', Edwards did not discuss possible health effects of contaminated meat or radioactivity in the environment. Chernobyl was an opportunity to criticize the government and address Edwards' primary nuclear concern: Sellafield. He argued that Sellafield not Chernobyl was responsible for the high levels of radioactivity in Cumbria. Thus, the article openly

questioned the official narrative of Chernobyl effects in Britain, however, far less radically than newspaper articles in the French case.

245. Another example of critical Chernobyl reporting is Robin McKie's 1991 article 'Britain's Deadly Chernobyl Legacy', addressing health effects in Britain: 'The milk was drunk by children, several dozen of whom are now expected to contract thyroid cancer over the next 30 years. This disturbing after-effect did not come to light until months after the disaster. Yet at the same time, Environment Secretary Kenneth Baker had claimed radiation was "nowhere near the levels at which there is any hazard to health."'
246. In this regard, it is no surprise that it was Edwards who, in July 2011, revealed the British government's attempt to influence media reports on Fukushima.
247. For Britain, this changed in the late 2000s as some Green Party activists, within the framework of the climate change debate, started to support nuclear power plants. This issue is addressed in the following chapter.
248. This phrasing is a simplification of reality as, since 1990, the 'British Greens' no longer exist. In 1990, the UK Green Party split into the Green Party of England and Wales, the Green Party in Northern Ireland and the Scottish Green Party.
249. However, already in 1999, the Green Party won a seat in the Scottish Parliament.
250. One of my British interview partners pointed out that the British deal with illness in their private sphere. A British person does not speak openly about their thyroid cancer in public. Therefore, even if there had been an increase in the number of illnesses, people would not necessarily know, and if their physician did not make the link with Chernobyl, they would not come up with this idea. Yet, I am uncertain whether this observation only exemplifies the 'keep calm and carry on' stereotype or if it is worth investigating. I think it would be counterproductive to open the frame of explanations for cultural stereotypes: to speak of 'national debates' makes sense as long as phenomena can be connected with aspects of the nation state, such as elite formation systems or national nuclear programmes. In case there is a link between supposedly British privacy regarding their illnesses and the National Health System or British labour rights, it might, however, indeed make sense to investigate this aspect in relation to the Chernobyl debate.

CHAPTER 3

2006

The Chernobyl 'Renaissance' within the 'Nuclear Renaissance'

The Transnational Chernobyl Debate

An insight into the transnational Chernobyl debate is crucial for understanding the national Chernobyl debates at the time of the accident's twentieth anniversary. In 2006, the Chernobyl Forum's report led to protests and profound critique from anti-nuclear and human rights activists. According to the Chernobyl Forum, the victims of Chernobyl radiation are a few firefighters who died of acute radiation syndrome, children who died from thyroid cancer and some potential 4,000 fatal cancer cases among the 'liquidators', evacuees and people in the highly contaminated area. The health effects on the five million residents of the other 'contaminated' areas (the report mostly put the word contaminated in inverted commas) are even more speculative as radiation-related deaths are expected to be less than one per cent of the normal cancer-induced mortality rate.[1] The report rejected that radioactive exposure from Chernobyl could cause DNA mutations that would affect future generations.[2] It acknowledged the increased incidence of many illnesses in these regions, however, it attributed these to increased screening and detection, improved recording methods and to mental health problems:

> The mental health impact of Chernobyl is the largest public health problem unleashed by the accident to date. Psychological distress arising from the accident and its aftermath has had a profound impact on individual and community behaviour. Populations in the affected areas exhibit strongly negative attitudes in self-assessments of health and wellbeing and a strong sense of lack of control over their own lives. Associated with these perceptions is an exaggerated sense of the dangers to health of exposure to radiation. The affected populations exhibit a widespread belief that exposed people are in some way condemned to a shorter life expectancy. Such fatalism is also linked to a loss of initiative to solve the problems of sustaining an income and to dependency on assistance from the state.

> Anxiety over the effects of radiation on health shows no sign of diminishing. Indeed, it may even be spreading beyond the affected areas into a wide section of the population. Parents may be transferring their anxiety to their children through example and excessively protective care.
>
> Yet while attributing a wide variety of medical complaints to Chernobyl, many residents of the affected areas neglect the role of personal behaviour in maintaining health This applies not only to radiation risks such as the consumption of mushrooms and berries from contaminated forests, but also to areas where individual behaviour is decisive, such as misuse of alcohol and tobacco.[3]

In this regard, the Chernobyl Forum reasoned that 'the most pressing health concerns for the affected areas thus lie in poor diet and lifestyle factors such as alcohol and tobacco use, as well as poverty and limited access to health care'.[4] The report[5] advocated resettling the regions evacuated in the late 1980s and phasing out any 'Chernobyl-related benefits and privileges'. The Chernobyl Forum considered potential health effects of low-level radiation negligible and therefore argued that 'the majority of the "contaminated" territories are now safe for settlement and economic activity'.[6] This report was presented to the public as state-of-the-art scientific research on Chernobyl, as summary of the international nuclear-scientific community's views. Anti-nuclear activists, however, saw it as the manifestation of the nuclear lobby's attempt to downplay the health effects of Chernobyl and of low-level radiation in general. They argued that many researchers' studies which did not strictly conform to this 'IAEA-authorized' narrative were simply ignored. These activists turned in particular to the works of four counter voices – Yuri Bandazhevsky, Vassily Nesterenko, Alexey Nesterenko and Alexey Yablokov – to corroborate their claim that the health effects of Chernobyl were much worse than this official evaluation claimed.

Yuri Bandazhevsky

Yuri Bandazhevsky is a Belarusian doctor who worked in one of his country's most heavily contaminated regions. On his own initiative, he carried out research on the relationship between the constant exposure of children to radionuclides in their food and their many illnesses, in particular respiratory and cardiovascular diseases. In 1999, Bandazhevsky was arrested on charges of corruption. Global protests against his arrest ensued, both by anti-nuclear as well as human rights organizations such as Amnesty International. In 2005, Bandazhevsky was finally released on parole. He went to France, where he planned to continue his research in cooperation with the CRIIRAD[7] and a stipend provided by the Conseil régional d'Auvergne. The CRIIRAD campaign made Bandazhevsky even more widely known in France, leading to official support in the form of several French cities granting him honorary citizenship. Well-known French TV presenter and journalist Noël Mamère, who stepped away from the media to pursue a political career with Les Verts – in his function as mayor of Bègles – conferred honorary citizenship on Bandazhevsky in 2009. Mamère had been involved in a longstanding legal battle with Pierre Pellerin and therefore had very personal close ties to the Chernobyl debate.[8]

Bandazhevsky's work was already known in France before his connection further strengthened when he moved there after his release from prison. Bandazhevsky's study results and imprisonment were prominent arguments for the narratives presented by the anti-nuclear side of the French Chernobyl debate, particularly those of the CRIIRAD. Other actors in the French Chernobyl debate also spread information about his work and what had happened to him. In 2004, Frédérick Lemarchand and Guillaume Grandazzi dedicated two articles in their book *Les silences de Tchernobyl* to Bandazhevsky and his work, one of which was by Bandazhevsky's wife Galina.[9] References to Bandazhevsky's work underpinned Lemarchand and Grandazzi's argument that in the most affected areas, the health effects of the radioactive fallout were much stronger than proclaimed by international expert groups.

Similarly, references to Bandazhevsky and his work featured in almost every publication written by critical voices at the time of Chernobyl's twentieth anniversary. For instance, Wladimir Tchertkoff elaborated on Bandazhevsky's (and Nesterenko's) work for his account of the 'global campaign for the denial of the Chernobyl victims',[10] and Stéphane Lhomme referred to Bandazhevsky when contemplating the role of the French nuclear sector in preventing Belarusian physicians from publishing their work on the 'true' impact of Chernobyl: 'without surprise, this unjustified imprisonment does not keep the French from collaborating with the Belarusian powers'.[11] To show his support, Stéphane Lhomme was also present when Noël Mamère made Bandazhevsky an honorary citizen of Bègles. Many other French authors referred to Bandazhevsky in their 2006 publications: In his book *Atomic Park: À la recherche des victimes du nucléaire* (Atomic park: in search of the nuclear victims) Jean-Phillipe Desbordes presented the trial against Bandazhevsky as an archetypal denial of the victims of the nuclear age.[12] In her graphic novel *Tchernobyl mon amour* (Chernobyl my love), Chantal Montellier included a reference to Bandazhevsky: at one point, the protagonist in the graphic novel – the young journalist Chris – thinks of the people who fought to reveal the truth about Chernobyl's health effects. Her speech balloon says: 'It was not only the fourth reactor that was choked, they also wanted to choke the truth. Those who spoke about the consequences of the catastrophe, such as the doctor and scientist

Yuri Bandazhevsky or the writer Svetlana Alexievich with her extraordinary book "La Supplication" were condemned to prison, to relegation and exile'.[13] On the same page, Montellier inserted hand-drawn portraits of Alexievich and Bandazhevsky. Alexievich's speech balloon consists of a quote from *La supplication*, and Bandazhevsky says: 'They arrested me for a ridiculous story of a bribe. But in reality it is my research that disturbs them'.[14]

But Bandazhevsky also contributed actively to the French Chernobyl debate. On the twentieth anniversary, his book *La philosophie de ma vie: journal de prison* (My life philosophy: prison diary) was published in France.[15] This was not a scholarly work but a personal log from his time in prison, along with scientific papers and an autobiography. The book's preface was written by Bandazhevsky's wife, Galina. She made it clear that their battle was not just about the health effects of Chernobyl, but also the threats posed by the nuclear industry in general: 'I hope that this book sparks the interest of a large group of readers sensitive to the problems of Chernobyl. That it will help them to understand the people who fight against the nuclear menace that weighs on our planet'.[16] The summary on the back cover asserted that Bandazhevsky's work was 'the scientific proof of the radioactive contamination of Chernobyl'.[17] For those in France who had long been critical of the official evaluations of Chernobyl, Bandazhevsky's book was precisely this: the scientific proof corroborating their criticism. The book was extensively covered in the media reporting for Chernobyl's twentieth anniversary: many newspapers published reviews or listed it as suggested reading, and France Culture broadcast an interview with Bandazhevsky. But *La philosophie de ma vie* was not the only book written by Bandazhevsky that appeared in France; *Tchernobyl 25 ans après* (Chernobyl 25 years on),[18] *Les conséquences de Tchernobyl sur la natalité* (The consequences of Chernobyl for the birth rate)[19] and *Les conséquences de Tchernobyl sur la santé* (The consequences of Chernobyl for health)[20] soon followed.

On the other side of the Channel, however, neither Yuri Bandazhevsky's work nor his imprisonment played a role in the British Chernobyl debate. The only account of Bandazhevsky's work I could find in my research on the British case was one newspaper article in 2006. Furthermore, his books have not been translated into English. The very different way his work is received in these two national contexts is apparent when consulting the Wikipedia articles on him: the French article has always been much longer and detailed than the English one.

Vassily Nesterenko, Alexey Nesterenko and Alexey Yablokov

Alongside Yuri Bandazhevsky, another Belarusian doctor, Vassily Nesterenko, considerably influenced the debate on Chernobyl's health effects. Nesterenko, who died in 2008, was the founder of the private Belarusian radiation protection institute Belrad. Belrad has, among other things, provided citizens in the most affected areas with facilities to measure the radioactivity in their food, and training in how to reduce their exposure. It also developed the highly contested pectin pills to reduce radioactivity levels in children. In France, mainly the group ETB promoted Belrad's and Nesterenko's work.[21] It is therefore not surprising that Wladimir Tchertkoff's *Le crime de Tchernobyl* included an account on Nesterenko, given that Tchertkoff was a founding member of the ETB. Beyond the inner circle of people linked to the ETB, however, Nesterenko has been far less visible in the French Chernobyl debate than Bandazhevsky. Nonetheless, one of Nesterenko's articles was included in the anthology *Les silences de Tchernobyl* edited by Frédérick Lemarchand and Guillaume Grandazzi. The article's main premise was that the Chernobyl accident had almost caused a nuclear reaction similar to an atomic bomb and thus could have rendered all of Europe uninhabitable.[22] Aside from this one, no further articles by Nesterenko have been published in French.

Throughout the English-speaking world, however, Vassily Nesterenko's and his son Alexey's publications are widely known. One book in particular, published in 2009 in collaboration with Alexey Yablokov[23] has become an important reference in the transnational Chernobyl debate – for the anti-nuclear as well as the pro-nuclear side.[24] For the anti-nuclear side, *Chernobyl: Consequences of the Catastrophe for People and the Environment* – is the long overdue scientific proof that the official reports on Chernobyl's impact and their explanatory pattern of radiophobia are erroneous and that various and severe health effects of the radioactive fallout are indeed observable. The pro-nuclear side, however, thinks that this publication just proves that no valid scientific argument can underpin the apocalyptic Chernobyl narratives. As this publication lies outside the timeframe of my book, I do not discuss this international controversy.[25] However, as the Chernobyl fallout in Britain figures prominently in this book, I want to shortly highlight this interesting detail. On several occasions, Britain was used as example of an affected region in Western Europe. Hardly surprising considering Yablokov and Chris Busby had jointly edited the book *Chernobyl 20 Years on* in 2006. Consequently, the Yablokov-Nesterenko report dedicated quite a bit of space to Busby's studies. It stated that 'in Wales, one of the regions most heavily contaminated by Chernobyl fallout, abnormally low birth weights (less than 1,500 g) were noted in 1986–87'.[26] It also referred to a study by Busby and Scott Cato that showed increased leukaemia in infants in Wales and Scotland, and quoted Busby: 'a significant increase in perinatal mortality occurred in March 1987, some 10 months after the catastrophe in the three most contaminated counties of England and Wales: Cumbria, Clwyd and Gwynedd'.[27] Alongside Busby's work, the report included another study on Chernobyl's health effects in Britain, this time in order to provide a perspective on the vast geographical impact of the fallout: *Thyroid Cancer in Children and Young Adults in the North of England: Is Increasing Incidence Related to the Chernobyl Accident?*, published in 2001 by Cotterill, Pearce and Parker in the *European Journal of Cancer*.[28] The findings of this research were as follows:

> Regression models showed a significant increase in the incidence of thyroid cancer after the Chernobyl accident (P=0.002). In Cumbria, the area receiving the heaviest fallout in the UK, the increase in incidence was much greater (rate ratio 12.19, 95% CI 1.5–101.2). These temporal and spatial changes in incidence are consistent with a causal association with the Chernobyl accident although a greater effect in the younger rather than the older age group would have been anticipated. However, factors including improvements in ascertainment and earlier detection of tumours may also have contributed to the increasing incidence.[29]

Interestingly, this study seems to have played a more important role in the transnational Chernobyl debate than in the British one. I did not come across any reference to this study in newspaper reports or in British Chernobyl literature. Furthermore: whilst Britain figured prominently in the Yablokov-Nesterenko report, the report itself was barely received in the British public discourse, although this slightly changed after the events at Fukushima.[30]

Chernobyl became a more prominent topic on transnational agendas at the time of its twentieth anniversary;[31] we will now look at how the national debates in France and Britain differed in 2006, starting by observing the media's approach to Chernobyl in this period.

Media Reporting on the Accident's Twentieth Anniversary

British newspaper editors and journalists did not consider the twentieth anniversary of Chernobyl a major news event.[32] Although significantly more items were published than in 1996,[33] the coverage remains modest compared to France. The French media presented the anniversary as a kind of an event in itself, with huge coverage throughout the second half of April 2006. French radio and TV stations dedicated substantial broadcasting time to Chernobyl. Most national newspapers started reporting on Chernobyl in mid-April already and published several pages-long supplements around 26 April, addressing various aspects of the Chernobyl accident: first of all the reasons why the accident happened, the current situation in the highly affected areas in Eastern Europe and the debate about the fallout's health effects in France. These supplements allocated extensive space to critical voices, who questioned the official narrative of the affected areas in Eastern Europe as well as the consequences of the accident in France.[34] Special attention was paid to interpretations by Guillaume Grandazzi, Frédérick Lemarchand, Galia Ackerman, Jean-Pierre Dupuy and Wladimir Tchertkoff; whose publications were often recommended to readers seeking further information on this topic. Thus, the French media coverage of Chernobyl's twentieth anniversary was strongly marked by an apocalyptic Chernobyl narrative, and at the same time it prominently addressed the issue of personal affectedness by recalling the *affaire Tchernobyl*.

Whereas personal affectedness had dominated the Chernobyl debate in France since the mid-1990s, in Britain this issue was only rediscovered at the time of the accident's twentieth anniversary. After having disappeared from the newspapers, the continued restrictions on sheep farms resurfaced in 2006. On 13 April, the *Guardian* published an article entitled 'Sheep Farms under Curbs See no End to Chernobyl Fallout'.[35] This article provided the latest news on the farms that were still subject to the restrictions and reported: 'now there are 359 in Wales, 9 in the Lake District and 10 in Scotland, involving 200,600 sheep.'[36] 'No one knows when the restrictions will end'. What makes this article particularly interesting is how it presented the persistence of the restrictions as though it was a new development: 'Hundreds of British sheep farms still blighted by the effects of radioactive fall-out after the world's worst nuclear accident, two decades ago at Chernobyl, will have to follow strict safety measures for years to come, it emerged yesterday'. Moreover, the article implied that it was not the journalists' non-interest that had led to this lack of coverage and ultimate disappearance from the public arena, but rather because 'few farmers wish to talk about their experiences, being worried about scaring consumers'.[37] However, the article made it clear that there was no need to worry since 'the [food standards] agency says the controls have effectively protected public food safety'. Not only was there nothing to worry about in terms of personal affectedness, the article also presented the general health effects of Chernobyl in an un-apocalyptic manner, ending with the statement: 'Officially, fewer than 50 people have so far perished as a result of Chernobyl, according to a study last year by the IAEA, the UN and the WHO'. Aside from this arguable interpretation of the Chernobyl death toll – which excludes the probabilistic cancer deaths assessed in this study – a closer look at this article reveals the deteriorating state of general public knowledge on the British nuclear fleet: when quoting a sheep farmer from Wales that the lasting restrictions 'make you think twice about living in the shadow of a nuclear power station, if they ever opened another one here', the author felt obliged to added: 'when Chernobyl happened there was a nuclear plant operating just a few miles south of the farm at Trawsfynydd' – an explanation that had not been necessary in the coverage of sheep farm restrictions in Wales in the 1990s, since the location of Trawsfynydd was then considered common knowledge.

A more general look at British newspaper reporting on Chernobyl's twentieth anniversary reveals more interesting findings. Similarly to the French reporting, British regional as well as national newspapers addressed a wide range of topics connected to the accident: from life stories of people in the most affected regions in Eastern Europe,[38] to the 'liquidators'[39] and the state of the wildlife in the restricted area around the plant.[40] Yet, there is a very important difference in their priorities: whereas in France, the *affaire Tchernobyl* – often paraphrased as 'the cloud that stopped at the border' – and the debate on the health effects of the fallout in France were a main theme of the reports, personal affectedness still only had a marginal place in British coverage. As in 1996, regional newspapers in particular highlighted the activities of local Chernobyl solidarity movement groups. The focus remained on the life stories of people actively involved in these groups as well as the many children hosted by them in Britain over the years.[41] These stories were often linked to fundraising drives in support of the Chernobyl aid groups, primarily the biggest associations, the CCLL, CCP (UK) and CCI.[42]

Although personal affectedness still did not play a major role in British news reporting, the topic was broached in some articles. And most interestingly, at least one journalist began to speak of the health effects caused by the radioactive fallout as a scientific finding and no longer as rumours or speculations – as with the Benbecula cancer story in 1996. On 23 April 2006, the national quality newspaper the *Independent on Sunday* published an article entitled 'Chernobyl "Still Causing Cancer in British Children"' which opened with the statement: 'More than a third of Britain is still contaminated by radioactivity from the Chernobyl disaster two decades ago, and children are getting cancer as a result'.[43] In this article, British personal affectedness had drawn up alongside French personal affectedness: the Chernobyl fallout had caused thyroid cancer in the country's children. Remarkably, this article also mirrored the positions taken during the French *affaire Tchernobyl* in how it addressed the issue of a deliberate cover-up by the government, albeit in a more subtle way compared to the intensity of the accusations directed at French officials:

> Scientists have found rates of thyroid cancer in children in Cumbria, the worst-affected part of England, rose 12-fold after the catastrophe – and blame fallout from the radioactive cloud that spread from the stricken reactor. This confounds government assurances at the time that the radiation in Britain was 'nowhere near the levels at which there is any hazard to health'.

This article, however, lies at the far extreme of what was said in newspapers about Chernobyl's health effects in Britain on the twentieth anniversary, and therefore in no way represents overall British media coverage in 2006.

Yet, the regional Liverpool-based newspaper the *Daily Post* also picked up on the topic of Chernobyl related illnesses in Britain, and on 26 April 2006 printed the article 'Wales: Did Chernobyl Disaster Cause my Cancer?'[44] Here, the focus was not increased cancer in children, but – like in the case of Benbecula – an increase in cancers among farmworkers. As with Benbecula, this personal affectedness was presented as a speculation: 'People in rural northwest Wales are worried cancer levels are higher than the national average. But they are also keen to avoid scare-mongering in case it damages tourism and farming'. Thus, there is a parallel between France and Britain insofar as in both countries it was discussed whether an evaluation of the true health effects of the fallout was suppressed for fear of negative economic repercussions. In France, this was an argument forwarded by 'Chernobyl victims' accusing the government of having put their health at risk in order to avoid negative effects on the French economy. In Britain, however, the people in the affected regions did not want the topic of potential fallout health effects publicized, for fear of the impact on their local economy. Taking this argument at face value means that British personal affectedness in terms of health effects was perceived as an individualized threat, and eventual disclosure was seen as a risk that things would change for the worse. Conversely, in France, the 'Chernobyl victims' poured all their energy into obtaining disclosure, not least because it promised monetary compensation and was therefore perceived as a change for the better.

Relating this far extreme of reported personal affectedness to general British newspaper coverage of Chernobyl's twentieth anniversary, it is important to stress that Chernobyl was typically discussed in British newspapers with relatively neutral remarks on the fallout of 1986 and the sheep farm restrictions, without any mention of potential health effects. The articles mostly presented Chernobyl as an event that had just happened. The rainfall and geographic constitution of the country were blamed for the problems with the fallout in Britain, not the civil use of nuclear energy. An article in the *Times* of 26 April 2006 can be considered typical of this common narrative:

> High-altitude winds spread the cloud of radiation over northern Europe, and on May 2, 1986, the fallout passed over Britain. … The toxic cloud would have caused no harm had the weather stayed dry, but local rainstorms washed down the contamination over a swath of Cumbria, Wales, Scotland and Northern Ireland. Radioactive caesium-137 contaminated the ground, and would have been bound up relatively safely in most soils. But the peat in upland areas kept the substance in solution and it became absorbed by grasses that were grazed by sheep. More than a million sheep were contaminated and emergency orders were imposed to prevent their sale. The legacy

> of that disaster still lingers in Britain. Although radiation levels have fallen over the years, some hills are still so contaminated that 375 farms and more than 200,000 sheep remain embargoed. No sheep can be moved out of affected areas without a special licence, the flocks are scanned regularly for radioactivity, and contaminated animals banned from markets.[45]

The fact that the account ends with the observation that the radionuclides had been taken up by the sheep, can be considered a statement in itself. There is no mention of potential effects of the radionuclides entering the human food-chain. People are not even mentioned in this article. The account somehow gives the impression that the only creatures that take up radionuclides are sheep. In this regard, it is very interesting to look closer at the above-cited article of 26 April 2006 in the *Daily Post*. This article also broached the issue of sheep farm restrictions, but with an added detail that was quite unique for British media reporting on the twentieth anniversary:

> Sheep were found to be contaminated and each year around 103,000 lambs are still tested for radioactivity before slaughter. A further 210,000 ewes and lambs are paint marked and licensed before being moved to grazing outside restricted areas. Because of the high level of monitoring, Welsh farmers say their lambs are the safest in the world. But they also claim animals have received more attention than humans.[46]

Another aspect of perceived personal affectedness is the way risks linked to Chernobyl are placed in relation to the national nuclear fleet or even other hazards. The article in the *Daily Post* offers yet another interesting perspective. The author quoted a farmworker saying: 'We've lived through BSE and foot-and-and-mouth and we don't want any more health scares' to which the journalist added: 'his farm is also down-wind of the decommissioned nuclear power station at Trawsfynydd'.[47] This tendency to relate Chernobyl to British-made hazards, especially the threat of national nuclear plants, has been an inherent element of the British Chernobyl debate from the very beginning and also showed up in the 2006 press reports. Another example is an article in the regional newspaper *Wales on Sunday* on 23 April 2006. It quoted Dr Keith Baverstock as saying 'that enough radioactive material to fill Albert Hall five times was being stored in "very much less than ideal conditions" at British power stations, including Wylfa on Anglesey'.[48] Yet, the London-based national newspaper the *Daily Telegraph* made a more 'traditional'[49] connection between the British nuclear enterprise and the Chernobyl accident: on 25 April 2006 it dedicated an entire article to 'Windscale Fallout from 1957 Reactor Fire Still Affecting Cumbria'.[50]

Public Authorities

In France, the twentieth anniversary of Chernobyl was not just widely covered by the media, the public authorities also published extensively on the accident. The CEA used the occasion to communicate its assessment of the accident to a broader audience.[51] It launched a comprehensive website with information on Chernobyl.[52] This website included a list of hyperlinks to (international) organizations involved in evaluating the situation in the most affected areas, as well as a general overview. This overview spoke of the accident's causes – faulty reactor design and human failure on the part of the plant workers – including a statement on the Chernobyl 'cloud' passing over Europe and France, which led to a multitude of studies investigating the impact of the fallout. This general account, however, did not elaborate on the results of these studies. Instead, the website posted a position paper by Bernard Bigot.[53] Bigot, Haut-Commissaire à l'Energie Atomique (High Commissioner for Nuclear Energy) had already drafted this paper in 2005, to provide a precise statement on the health-related consequences in France, which people, and especially French politicians, could refer to in the upcoming debate for the twentieth anniversary.[54] Bigot's account of Chernobyl opened with a description of the accident's causes, but very soon switched to its main topic, the *affaire Tchernobyl*. He stressed that a wide range of data had been available to the public in May 1986, derived from the substantial number of measurements taken by the French public authorities.[55] Bigot then addressed the various maps showing the radioactive deposits in France. He explained that the differences between the maps were due to the different models used to create them. As for the health effects, he stated that the only possibility was thyroid cancer; however, none of the cancer registers (including the Provence-Alpes-Côte d'Azur and Corse regions) had shown any increase. Even if additional cases of thyroid cancer occurred, this number would be so low that it would not be possible to detect these incidences.[56] Thus, he stated, 'concerning the risk that resulted from Chernobyl, we are confronted today with a finding that is purely statistical'.[57] Aside from the debate on Chernobyl health effects in France, Bigot's paper focussed on the criticism directed at the French crisis management. From his point of view, there was no reason to question the public authorities, given that their action had been absolutely proportional to the situation.[58] Even if today's calculations would verify additional cases of thyroid cancer in France, any countermeasures would have been disproportional to the increased fear and the negative economic consequences of these countermeasures.[59] Bigot did not limit his discussion of Chernobyl's impact to France. He also provided an evaluation of Chernobyl health effects in Eastern Europe, stating that the only effect that could actually be observed was an increase in thyroid cancer in children. Regarding other cancers, leukaemia and general mortality rate, no effects were yet discernable.[60] Bigot's fundamental stance was to emphasize the continuing excellent work by French nuclear institutions and experts. In his opinion, there was no reason to criticize any of the decisions or evaluations made in 1986 or in 2006. Hence, Bigot appealed to the French government to finally put an end to the misleading debate on Chernobyl health effects in France.[61]

Like the CEA, the IRSN published an extensive dossier on Chernobyl on its website around the time of the twentieth anniversary.[62] This dossier[63] not only contained general reports the IPSN and later its successor IRSN had published every year since 1996, but also information and maps on the radioactive fallout in France. The website included a detailed bibliography, listing publications by a number of critical voices who attributed either more or less importance to the Chernobyl impact than the IRSN. In its account on the Chernobyl death toll, the IRSN noted the factors that made it so difficult to provide a concrete number: including the lack of knowledge on the health effects of low-level radiation and the difficulties in conducting representative studies. The IRSN thus concluded that 'generally speaking, the prediction of the number of deaths caused by the received doses is tarnished by severe incertitude'.[64] To demonstrate the range of this uncertainty, the IRSN quoted its analogous British institution, the NRPB, which had calculated the death toll as anywhere between 4,200 and 80,000. Regarding the health effects in France, the IRSN pointed to the study the IPSN and InVS had conducted in 2000 and repeated that, according to the models applied, the number of additional incidences of thyroid cancer in children was still within the natural range, and therefore any cancers caused by the accident would not have been detectable.[65] To a large degree, the IRSN's Chernobyl dossier was dedicated to nuclear safety and examining how the accident had challenged this field. The IRSN stressed that the studies in the most affected regions of Eastern Europe were of paramount importance for gaining knowledge about living conditions in contaminated areas.[66] Thus, the IRSN was concerned about the health situation in Eastern Europe and believed that the consequences of the accident had not yet been fully evaluated. However, the IRSN saw no reason for a debate about French personal affectedness. In the IRSN's view, even if a health impact existed in France, it would be far too small to observe.

In direct relation to the twentieth anniversary, albeit with a slight delay, Philippe Renaud, Didier Champion and Jean Brenot published *Les retombées radioactives de l'accident de Tchernobyl sur le territoire français*[67] (The radioactive fallout from the Chernobyl accident on French territory). The book came out in 2007 and was an updated version of the 1999 book by Philippe Renaud et al., *Les retombées en France de l'accident de Tchernobyl*. As stated in the preface, this new book was part of the IRSN's efforts to draw a balance regarding the Chernobyl health effects on this important anniversary.[68] In the chapter

'Estimation des risques sanitaires en France' (Estimation of the sanitary risks in France), the authors stated that the calculated doses justified an evaluation of the health effects in France, particularly regarding possible thyroid cancers.[69] However, the evaluation results they presented confirmed there was no reason to worry about the Chernobyl health effects in France:

> The results show that, according to the model, the excesses of the estimated cases are inferior or comparable to the uncertainties regarding the estimates on the number of 'spontaneous' cancers: thus, these excesses would be hard to detect from an epidemiological point of view. Considering the limits of the indicated method and the uncertainties regarding the risks associated with low doses, the real excess of thyroid cancer, on the level of the considered doses, might actually be zero.[70]

Concerning 'the lessons learned in France on the management of nuclear crises', the authors stated that their evaluation was not intended to question the appropriateness of the French public authorities' actions in 1986. On the contrary, rather than looking back to ask what should and could have been done in the case of Chernobyl, it was more important to ask what should and could be done in the event of another significant nuclear accident.[71]

The InVS also used the twentieth anniversary of Chernobyl to present its evaluation of the health effects in France to the public at large.[72] To this end it published the pamphlet *Surveillance sanitaire en France en lien avec l'accident de Tchernobyl – Bilan actualisé sur les cancers thyroïdiens et études épidémiologiques en cours en 2006*[73] (Health observation in France in connection with the Chernobyl accident – updated conclusion on thyroid cancers and epidemiological studies under way in 2006). Since 2000, the InVS had investigated the possible increase of thyroid cancers in France by compiling a broad range of cancer registers and local epidemiological studies. The InVS presented its results in 2006, not including the outcome of the investigations in Corsica as this study had not yet been completed.[74] InVS's further investigations confirmed the initial results published in 2001 and 2003, namely that 'the observed increase in incidences of thyroid cancers is longstanding, important, and continued in time, having started before the Chernobyl accident. … The lowest rates were observed in the counties of Alsace, which had been the most exposed to the accident's fallout'.[75] The studies revealed a slight increase in already very rare incidences of thyroid cancer in children. But according to the InVS, this observed increase, just like the general increase in thyroid cancers, was not linked to Chernobyl, but to the improved registration of these cases over time.[76] However, considering the causes of thyroid cancer were so little known or understood, the InVS announced it would be investigating this topic further, particularly the potential role of ionizing radiation in developing these cancers in children.[77] In this regard, the aspect of French personal affectedness figured prominently in the InVS report.

The InVS pamphlet did not just provide an evaluation of the health impact in France, it also spoke of the health impact in the worst affected areas of Eastern Europe: since 1990, there had been an epidemic of thyroid cancers in children, but there was no evidence of a significant increase in other cancers in the three most contaminated countries.[78] The way the InVS evaluated Chernobyl in this pamphlet can be directly linked to its position regarding the health impact of low-level radiation: 'There is not any epidemiological study that has been able to show that incidences of cancer are linked to low-level exposure'.[79]

British public authorities acted quite differently. In 2005, the NRPB became the Radiation Protection Division of the Health Protection Agency (HPA). Thus, in 2006, the task of communicating information on Chernobyl fell to the HPA. However, there are no publications in Britain that are comparable to those released by the IRSN and the InVS in France.[80] On the twentieth anniversary, the HPA only published a two-page press release entitled 'The Consequences of the Chernobyl Nuclear Accident'.[81] The numbers presented were taken from the articles written by Elisabeth Cardis et al. in 1996 and 2006.[82] To give a better overview of these numbers, the HPA reproduced the summary table from one of the articles, showing the 'Mean cumulative whole body dose', the 'Predicted excess number of cancer deaths', and the 'Predicted % of cancer deaths due to radiation in the population'. The HPA included only two more references: the 2000 UNSCEAR report and the 2006 WHO/Chernobyl Forum report (of which Cardis' findings were an essential part). The press release was only three paragraphs, and as it is one of the very few Chernobyl statements released by British public authorities after the mid-1990s, it deserves quoting here in full:

> The 20th anniversary of the Chernobyl accident on 26th April has produced a number of papers, articles and claims about the health effects of the accident both here in the UK and in Eastern Europe. The estimates of the number of health effects caused by radiation exposure from the accident vary widely, from a few thousand to hundreds of thousands of deaths.
>
> Three people died immediately as a result of the accident and a further twenty-eight died within a few weeks as a direct result of acute radiation doses. They were staff working at the Chernobyl nuclear power station at the time and staff from the emergency services, particularly the fire service. Nineteen more of these emergency workers died during the period 1987 to 2004 from various causes. There is also an increase in the incidence of thyroid cancer in people who were children in 1986, including those in utero at the time of the accident. At present over 4000 cases of thyroid cancer have arisen in

> Belarus, the Ukraine and parts of Russia, most of which can be attributed to exposure of the thyroid gland by radioactive iodine isotopes from Chernobyl. This condition is fatal in only about 1 % of cases but it is nevertheless a serious health effect; people affected need to take medication for the rest of their lives.
>
> Predictions of increases in the incidence of cancer in general, and of other illnesses that might have resulted from exposures to radionuclides from Chernobyl, are subject to large uncertainties and can therefore be contentious. These uncertainties are at their greatest when attempting to estimate the number of excess cancer cases attributable to very low radiation doses received by very large numbers of people. A wide range of estimates have been reported in recent weeks using various risk factors and differing methods of calculation. The most reliable recent evidence comes from Elisabeth Cardis and colleagues published in the *International Journal of Cancer*. We reproduce a summary table of their predictions below and the uncertainties inherent in making such predictions are discussed in detail in the paper and in a recent review.

The figure presented as the 'Predicted excess numbers of cancer deaths' for the whole of Europe was 16,000. In relation to other calculations at the time of Chernobyl's twentieth anniversary, this is one of the lower estimates. The more interesting feature of this press release is that the HPA's mention of the Chernobyl fallout in Britain is marginal. Thus, personal affectedness was absolutely a non-issue for the British radiation protection authority. Regarding the third narrative element, anti-East European/anti-Soviet stereotypes, nothing can be derived from these few lines. However, another source, the official memoir of the NRPB published by the HPA one year after the twentieth anniversary, is interesting for this purpose.[83] In this institutional history of more than 300 pages, the author dedicated only three pages to Chernobyl, praising the 'excellent manner in which the scientific, administrative and public relations aspects of the event had been handled'.[84] The problems in Britain due to the fallout were omitted, and the author framed the Chernobyl impact in, for 2007, quite anachronistic Cold War rhetoric:

> Although the accident caused thirty deaths among the workers at the site and increased incidence of thyroid cancer among children, considerable disruption of the economy, and much distress in the affected areas, the people of the region were inured to suffering. They, or their parents, had experienced the horrors of collectivisation, famine, purges, invasion and the Gulag system. If there was a positive outcome of the accident, it was at the political level in the USSR.[85]

Nuclear Power Industry

The British nuclear industry did not communicate on Chernobyl at the time of the twentieth anniversary. I did not find one single publication or even a press release from any of the actors in this cluster. Like in France, the British nuclear industry withdrew from the Chernobyl debate after the early 1990s. However, there is one very important difference between these two cases. Whereas in France, in the years leading up to the twentieth Chernobyl anniversary, the nuclear industry, in particular Areva, had received substantial political support, the British nuclear power industry had basically been dismantled with the privatization of the British energy sector – a privatization that resulted in the sale of the British nuclear power plants to the British branch of the EDF, EDF Energy. Thus, in 2006, the actors in the British nuclear power industry were very much linked to France.

It seems that after its publications on Chernobyl in the late 1980s and early 1990s, the EDF felt it had said everything there was to say on the topic. Similar to the actors in the British nuclear industry, the EDF's immediate communication revolved around the central statement that an accident like Chernobyl could not happen in its own power plants. Any possible links between the company's activities and the events at Chernobyl were rejected. Therefore, I hypothesize that the EDF did not further communicate on Chernobyl because the company simply did not consider itself part of the Chernobyl debate. The possibility that people would link the accident to the EDF's activities if it continued to release statements and participate in the Chernobyl debate was probably the company's reason for not issuing further statements. This policy to avoid any link between Chernobyl and French nuclear power plants probably also impacted the twentieth anniversary festivities at the Paluel site. Although this site opened in April 1986, the EDF celebrated its anniversary at the end of 2006. A celebration in April 2006 would probably have become a target for anti nuclear protests. The EDF's public relations communication is another example of how the company has actively distanced its affairs from the accident. The company history on the EDF website does not mention Chernobyl at all. EDF's decision not to discuss Chernobyl can be considered the logical consequence of how the events of 1986 were classified, namely as a Soviet accident. The EDF is proud of its nuclear power plants and openly advertises the source of its generated electricity; it has done so even more openly since the 'greening of the atom' within the climate change debate – strongly lobbied for by France[86] – discursively transformed nuclear reactors into mechanisms of nature conservation, if not of world salvation.

Similar reasoning probably underpins Areva's silence on the topic of Chernobyl. Neither Areva nor its precursors Framatome and Cogema published any statements on the accident. However, we can deduce Areva's internal interpretation of Chernobyl from another source: a book written by Areva's long-term chief executive Anne Lauvergeon and her coworker Michel-Hubert Jamard. Published in 2008, *La troisième révolution énergétique*[87] (The third energy revolution) lies outside the temporal boundaries of this book. However, because Lauvergeon and Jamard's account of Chernobyl is interesting, and the only Areva-linked statement on Chernobyl I came across in my research, I include it here. The space dedicated to Chernobyl in *La troisième révolution énergétique* is minimal; only four out of 200 pages. The title of this account, 'Une castrophe soviétique', already reveals which direction their argument would take: according to the authors, the accident was caused by the operators' action and the reactor design. But Lauvergeon and Jamard provided more than just their view of the events in 1986. They also assessed Chernobyl's health effects. They referred to the Chernobyl Forum's report, stating that it was 'the most profound study that has been conducted so far'.[88] They went on to state that there would be 4,000 additional deaths as a result of the accident, asserting that the 'relatively low level of doses received by the inhabitants of the affected areas' did not substantiate the presence of any links between the radioactive fallout and the increase in still-births, malformations, or the weak health of children in this region.[89] According to the Chernobyl Forum experts, the reason for this problematic health-related situation was people's mental health problems; lack of information had led to them being convinced they had a reduced life expectancy; 'and it is not the apocalyptic statements by anti-nuclear activists that can help these people to carry on with their lives'.[90] The authors also included an account on French personal affectedness: epidemiological studies had shown no correlation between an increase in thyroid cancer in a country like France, and Chernobyl.[91] Lauvergeon and Jamard's Chernobyl narrative is well summed up in a metaphor they added, clearly alluding to anti-Soviet stereotypes:

> In consideration of these facts, it becomes clear that the attempts to make Chernobyl the incarnation of the dangers implied in nuclear energy, are the result of ignorance – or of maliciousness. A little bit like if one were to condemn cars because of a deadly accident that had been provoked by a Trabant, the brakes of which were sabotaged and that had been driven at full speed on a mountain road by a driver who was determined to ignore traffic rules.[92]

Critical Voices

Unlike the EDF's strategy of distancing its company activities from Chernobyl, French anti-nuclear activists were very keen to make the connection between the accident and the French nuclear enterprise. The campaign protesting the construction of the first EPR in France – at the Flamanville site in Normandy – emphasized this connection. This project would be the first of an entire fleet of new nuclear reactors to be built in France; Flamanville 3 was also intended as a showcase to market this new reactor, particularly after problems with constructing the first ever EPR had been compromised in Finland. This setback had threatened to ruin the reputation of the new reactor design, which was supposed to become the new cash cow of French export. In 2005, the public authorities gave the go-ahead for the project in Flamanville, where two reactors already existed. In April 2006 – the same month as Chernobyl's twentieth anniversary – thousands of people gathered to protest against the EPR in Cherbourg.

In the years leading up to, including and following 2006, Greenpeace France focussed its anti-nuclear campaigns on the EPR. At the same time, this NGO was one of the main promoters of a public commemoration of Chernobyl's twentieth anniversary. For the occasion, Greenpeace published a pamphlet[93] and a CD[94] with songs written by various French artists on Chernobyl. Together with Fnac, Greenpeace also organized a photo exhibition in the Parisian shopping mall Forum des Halles, performances of the theatre play *Une autre voix solitaire* (a play based on *La supplication*) and round table discussions.[95]

Réseau sortir du nucléaire was the other key player to campaign against the EPR project in Flamanville. At the same time as mobilizing the anti-EPR demonstration in Cherbourg in mid-April, the network prominently raised the issue of Chernobyl. On its website, it published an extensive dossier on Chernobyl, focussing on the impact in France and the *affaire Tchernobyl*.[96] In early 2006, the network's spokesperson Stéphane Lhomme published his book *L'insécurité nucléaire: Bientôt un Tchernobyl en France*[97] (Nuclear insecurity: soon a Chernobyl in France). Although the accident was more of a metaphor than the theme of the book, the Chernobyl anniversary was the right occasion for publication.[98] Lhomme's primary concern was the risk that a similar accident could happen in France. In his opinion, Chernobyl was everything but a Soviet accident: Western nuclear experts had been familiar with the RBMK reactor design, however, they did everything to dismiss their complicity. Lhomme also pointed out that the human failure of plant operators could happen anywhere in the world. His description of the situation in the worst affected regions of Eastern Europe drew heavily on the narrative provided in *La supplication*. But Lhomme's aim was not to present a detailed account of the health situation of people living in these regions, but to reveal the overarching politics governing the management of these people's daily lives. From Lhomme's point of view, these people were being treated like laboratory animals in a huge experiment, clearly aiming 'to "prove" by all possible means that after all, the consequences of Chernobyl, and by deduction any other eventual nuclear accident, were just not that serious'.[99] In particular, the ETHOS project and the CORE programme[100] were intended as demonstrations of 'how to live happily in a contaminated area'.[101] Lhomme identified a clear promoter of this policy: the French *nucléocratie*. According to Lhomme, the French *nucléocratie* was using the contaminated areas around Chernobyl as a training ground to learn and prepare for such eventualities as an accident in a French nuclear power plant. At the international level, this way of managing and simultaneously benefitting from the situation in the most affected areas of Eastern Europe was made possible by the IAEA–WHO agreement and the concept of 'radiophobia'.[102] French nuclear policies and Chernobyl were tightly entwined in Lhomme's narrative: the *affaire Tchernobyl* demonstrated the lies and cover-ups the *nucléocratie* propagated regarding the dangers of the nuclear enterprise.[103] At the same time, he highlighted how the *nucléocratie* had dominated the evaluation and management of the situation in Eastern Europe.

Thus, for French anti-nuclear activists, promoting a public commemoration of Chernobyl on its twentieth anniversary served two purposes. On the one hand, Chernobyl was presented as proof that a serious accident could indeed happen and would have devastating consequences. On the other hand, by linking the management of the post-accident situation in France and Eastern Europe with the power structures of the French nuclear techno-political regime, the activists used Chernobyl to demonstrate the inherent inhumanity of French nuclear policies. Activists felt that there was an urgent need to unveil the real implications of these policies for the future of France, if not the rest of the world. After all, the French nuclear industry was about to reap the benefits of its international lobbying efforts to 'green the atom' and was preparing to sell the new EPR to the rest of the world. Therefore, within the context of the launch of the new build EPR project in Flamanville, the twentieth anniversary of Chernobyl proved to be the ideal opportunity for the activists to draw the wider public's attention to their cause. They contrasted the image of the clean and green energy of the future, the embodiment of the EPR marketing campaign, with the apocalyptic scenery in the worst affected regions of Eastern Europe. Combining this diverse imagery was their tactic to show that nuclear power was anything but a benign technology. The activists wanted people to think twice about whether they truly found it acceptable to be surrounded by nuclear plants in their country, and whether they really supported a policy that pushed for worldwide nuclear expansion.

The CRIIRAD also instrumentalized the twentieth anniversary of the accident to draw attention to its cause, calling to account the people who had covered up the 'true' impact of the Chernobyl fallout in France.[104] In early April, the CRIIRAD released *Tchernobyl – 20 ans après, les services officiels français persistent dans la censure et la désinformation*[105] (Chernobyl – 20 years later, the French authorities' persisting censorship and disinformation) and made this communiqué available on its website. The publication's main argument was its criticism of the IRSN: the SCPRI's successor would still not admit that the data published in 1986 had been falsified. Even worse, the same people who had lied to the French public twenty years ago were still in charge and the same mind-set that had made these lies possible persisted. 'In the event of a new accident occurring, one thing is certain: there will not be a single discordant voice inside the official expert institution, everybody will speak with one voice – and this voice will have the same intonations as in 1986!'[106] That the *affaire Tchernobyl*, even twenty years later, was still considered an ongoing struggle between the very same actors as in 1986, is well illustrated by a round table discussion that the CRIIRAD had attempted for 25 April 2006: Bella and Roger Belbéoch, as well as Roland Desbordes and Corinne Castanier of the CRIIRAD were supposed to confront Philippe Renaud, Georges Charpak, André Aurengo and André-Claude Lacoste.[107] The CRIIRAD's commitment to Chernobyl was not limited to resolving the issue of the *affaire Tchernobyl* and ensuring compensation for the health effects of the fallout in France. The organization also supported East European Chernobyl activists, in particular Yuri Bandazhevsky, and promoted their evaluations of the devastating Chernobyl health effects in Eastern Europe to the wider French public.

Advocates who challenged the official evaluations of the most affected areas in Eastern Europe also used the anniversary to increase awareness for their cause.[108] To this end, Frédérick Lemarchand and Guillaume Grandazzi published a revised and expanded edition of *Les silences de Tchernobyl*.[109] This time around, Galia Ackerman joined the team of editors, who also pursued individual projects in 2006. Galia Ackerman published her own book *Tchernobyl, retour sur un désastre*[110] (Chernobyl, return to a disaster) and contributed to the French edition of Igor Kostin's photography book on Chernobyl.[111] Guillaume Grandazzi, in turn, wrote an article for the German journal *Osteuropa*,[112] and Frédérick Lemarchand, together with others, edited a special issue for the journal *Ecologie & Politique*.[113] Among this group of editors was Jean-Pierre Dupuy, philosopher of science and author of the book *Pour un catastrophisme éclairé*[114] (In support of an enlightened catastrophism) who had become more involved with Chernobyl through his joint visit with Grandazzi and Lemarchand to the most affected areas. Dupuy published a book on Chernobyl on the twentieth anniversary as well: *Retour de Tchernobyl: journal d'un homme en colère*[115] (Return from Chernobyl: diary of an angry man).

The commonality of all of these publications lay in their emphasis on the devastating consequences the accident had had and would continue to have for the people living in the most contaminated areas of Eastern Europe. The authors vehemently criticized international attempts to reject any connection between the health situation and the radio-hygienic situation of the areas, and to dump the individuals of the contaminated areas with the responsibility of dealing with the consequences themselves. The sheer number of publications originating from the sphere of the 'Caen-Group' ensured that their narrative of Chernobyl would profoundly shape the media reporting on the twentieth anniversary. Interviews with the authors were printed and broadcast, their books were reviewed and figured prominently in reading recommendations, while Galia Ackerman ensured the broad and very critical coverage of Chernobyl on France Inter.

Undoubtedly, there were also people in Britain who used the twentieth anniversary to criticize the evaluation and management of Chernobyl's impact. Their criticism was particularly directed at the official IAEA and WHO reports published in late 2005 regarding the environmental and health impact of the accident. In response to these reports, MEP Rebecca Harms, commissioned a study to investigate the scientific basis of their evaluations, which was published in 2006 and became widely known as the TORCH-report (The Other Report on Chernobyl).[116] Although the authors, Ian Fairlie and David Sumner, are both British scientists, the report has never truly been perceived as a British publication, least of all in the UK. This is probably due to the fact that a wider British debate on Chernobyl never materialized during the twentieth anniversary. The few British activists who took part in the transnational Chernobyl debate did not collaborate in a concerted effort at a national level, like the French Chernobyl activists. The French activists received far more public attention thanks to the strong perception of personal affectedness, which had become an even more pressing political issue within the context of the EPR new build project. This public interest in turn generated enough interest in publishers, newspaper editors and television broadcasters for them to cover the topic. In Britain, however, the issue of nuclear energy policies had almost totally disappeared from the public discourse for years: for instance, the government's 2003 Energy White Paper had not even included nuclear energy in its considerations. Yet, 2006 was the year that put nuclear energy back on the agenda: Tony Blair's Energy Review, launched in January 2006, embraced the many voices that had lobbied to revive the British nuclear enterprise. Many of these voices had argued that nuclear power would be the perfect solution to reduce CO_2-emissions and achieve the ambitious aims the British government had set for itself to fight climate change. This turnaround in the Labour government's nuclear policies was consolidated in the years to come: in 2007, the EDF publicly expressed its willingness to operate new build reactors,[117] and – after intensive debates on the costs of the new build project and the legitimacy guaranteeing the EDF a fixed price per kW/h[118] – the government gave its approval in March 2013 for Hinkley Point C. Interestingly, Greenpeace UK's criticism against this decision mainly drew on an economic framing: 'Hinkley C fails every test – economic, consumer, and environmental. It will lock a generation of consumers into higher energy bills, via a strike price that's understood to be nearly double the current price of electricity, and it will distort energy policy by displacing newer, cleaner, technologies that are dropping dramatically in price.'[119]

But in 2006, this was all still very much perceived as the government's dreams for the future. British anti-nuclear power activists considered it very unlikely that the government would be able to pursue this goal. Thus, in 2006, there was no reason why the debate on the 'true' impact of Chernobyl should become a proxy war against national nuclear policies as had occurred in France. By 2006, Chernobyl was entirely distanced from British nuclear policies. However, the fact that some newspapers included personal affectedness in their twentieth anniversary reporting might be directly linked to this change in nuclear policies, since it put the newsworthy risks of nuclear power plants back on the agenda. Yet, in order to get statements on British personal affectedness, journalists basically had to turn to one individual campaigner: Chris Busby.

Individual Voices

Although Chris Busby's contribution to the Chernobyl debate on the twentieth anniversary was a cooperative project from a transnational perspective, I here treat his as an individual voice as his collaborators were from outside the British nuclear discourse. In 2006, together with Alexey Yablokov, Busby edited the book *Chernobyl 20 Years on*.[120] This compendium of articles was published by Busby's own environmental consultancy firm Green Audit. The collection investigated various aspects of Chernobyl's health effects and was mostly written by Eastern European scientists. According to the editors, the studies had been deliberately ignored by Western radiation experts because they openly contradicted the official narrative of Chernobyl's health impact, a narrative endorsed by the UNSCEAR and the WHO. The book was not just a forum for a Chernobyl counter narrative. The aim was to present evidence that the models evaluating the risk of radiation exposure – developed by the International Commission on Radiological Protection (ICRP) and based on studies of Hiroshima and Nagasaki survivors – were erroneous. This claim employed a rationale far beyond the case of Chernobyl and nuclear accidents. It questioned the risk assessment of nuclear power plants in general – again, not due to the fear of what could happen in the event of an accident, but because of the radioactive emissions from normally operating nuclear power plants. Although this publication addressed the Chernobyl accident, its primary aim was to highlight the need for a fundamental review of the risks associated with nuclear power in general. The editors explained Chernobyl's role in their argument: 'The Chernobyl accident and its appalling outcomes have given the human race the empirical evidence to test this theory [the ICRP model]'.[121] And they also openly addressed the wider political implications of their assessment:

> This book represents a landmark on the road to understanding the effects of low-dose chronic irradiation. The committee believes that these lessons should be borne in mind by policy makers who are, even now, discussing new investments in nuclear energy and ways in which historic and future radionuclide waste can be disposed of into the environment. The committee recommends this book to scientists and policymakers and concerned members of the public in the hope that the huge amount of work carried out by scientists publishing their results in Russian language journals and others studying the effects of the Chernobyl accident will influence their decisions in this important area of public health.[122]

This quote is especially interesting because of its clear reference to the historical context in which this book was published: 'policy makers who are, even now, discussing new investments in nuclear energy', or rather Tony Blair's government, among others. Thus, Busby, very much like anti-nuclear activists in France, used Chernobyl as an instrument to counter national nuclear policies and to influence public opinion with his evaluation of the accident.

The articles in the compendium were intended to leave no doubt in the reader's mind that Chernobyl had generated an apocalypse. The radiation released by the accident had caused changes in the genomes of 'all species, plants and animals and humans'.[123] The effects of these changes were the topic of the various articles compiled in this book. For example, Krysanov wrote about the higher radiosensitivity of mice living in the high irradiation zone for generations; Yablokov contributed an article on the health status (or rather sickness status) of the 'liquidators'; and the Nesterenkos elaborated on the various illnesses found in children and adults in the most affected areas of Belarus that they had been studying for years. But the apocalyptic narrative of this collection of reports most obviously surfaced in Rosalie Bertell's article on 'The Death Toll of the Chernobyl Accident'. Bertell, one of North America's most prominent anti-nuclear campaigners, proclaimed that a conservative estimate hovered between '899,600 to 1,787,000 in total'.[124]

The fact that this book – published by a British anti-nuclear campaigner in close cooperation with a Russian colleague – included papers by the Nesterenkos, Inge Schmitz-Feuerhake and Rosalie Bertell and directly addressed the health effects of Chernobyl from a global perspective, reveals to what extent the Chernobyl debate had become transnationally entangled by 2006. These well-connected transnational Chernobyl activists were now mobilized by Busby to underpin his claim that a re-evaluation of low-dose radiation effects was necessary. This was supposed to be the starting point for what he hoped would become a more generalized campaign against the nuclear turn the British government had just taken.

The way these two debates were connected is most visible in Busby's own contribution, in which British personal affectedness formed an integral part. In his article on 'Infant Leukaemia in Europe after Chernobyl and Its Significance for Radioprotection: A Meta-Analysis of Three Countries Including New Data from the UK', Busby argued that 'infant leukaemia rates had been increased following exposure in utero to radiation from the Chernobyl accident fallout in the UK'.[125] But the point he wanted to make was not so much about Chernobyl's impact in Britain. For Busby, this was just one stepping-stone among many in a general argument built around the health impact of nuclear power plants during their normal, accident-free operation. His line of argument aimed to prove that the low-level Chernobyl radiation to which people had been exposed led to serious health problems, implying that nuclear plants in general caused cancer in the people living near them. According to Busby, these health effects had already been proven by the child leukaemia cluster verified around Sellafield.[126] For him, there was only one possible conclusion: 'In the case of the Chernobyl infant leukaemias there is no alternative explanation apart from internal radiation exposure to largely the same isotopes as the nuclear site leukaemias. The significance of this result for radio protection is overwhelming'.[127] Busby used the research on the impact of Chernobyl in Britain as a means to a very specific end: the end of electricity generation through nuclear power. And like in his earlier publications, the health effects of Chernobyl in Britain served to illustrate the impact of the global nuclear enterprise.

This is an interesting difference compared to the focus underpinning the French Chernobyl debate. In France, the Chernobyl fallout has been perceived as something dangerous in itself, not necessarily related to the national nuclear enterprise. This setting was a result of the prominent role attributed to French personal affectedness in the debate on Chernobyl's impact. In Britain, however, the Chernobyl fallout was used as a means to expose the dangers of British nuclear installations. Investigations into the health effects of the Chernobyl fallout in Britain aimed to underpin arguments brought against the British and global nuclear industry and did not focus on individual compensation.

In France, various individual voices used Chernobyl's twentieth anniversary to disseminate their interpretations and narratives.[128] Among them was one actor whose work we have already discussed: Jean-Michel Jacquemin. In 2006, he released his new book *Tchernobyl 20 ans après: Cachez ce nuage que je ne saurais voir*[129] (Chernobyl 20 years on: Hide this cloud that I can't endure to look at). Creating with his title a reference to Molière's famous comedy *Tartuffe, or the Hypocrite*, the 400 pages that followed were basically a synthesis of his earlier publications. It emphasized the devastating health effects of Chernobyl in France and continuing official cover-up of the true impact. Jacquemin also contributed an article to the new edition of *Les silences de Tchernobyl.*[130] But, on the twentieth anniversary, also new individual actors emerged. For instance, Chantal Montellier wrote and illustrated a graphic novel on Chernobyl that not only addressed the issue of the health effects in Eastern Europe, but also broached the topic of the *affaire Tchernobyl.*[131] Raymond Micoulaut published a book on the secrecy policies surrounding the crisis management of the Chernobyl accident, in France as well as in the Soviet Union.[132] Moreover, for Chernobyl activists from Eastern Europe like Igor Kostin[133] and Yuri Bandazhevsky,[134] the anniversary was an opportunity to publish in France their accounts of the accident and its impact. It was also a convenient occasion to publish books on the nuclear enterprise in general. Thus, it is no coincidence that Jean-Philippe Desbordes' book *Atomic Park* and the French translation of Günther Anders' classic *Die atomare Bedrohung*[135] (The atomic threat) appeared in 2006. This variety and quantity of nuclear-related publications in 2006 demonstrate Chernobyl's central role in the French nuclear discourse.

Chernobyl Solidarity Movement

In France, the solidarity movement groups that organize recreational stays for children were almost invisible in 2006. They did not publish anything on their work, nor did the media dedicate much space to their activities. However, the Chernobyl interpretations of two founding members of the ETB, Galia Ackerman and Wladimir Tchertkoff, figured prominently in news reports. In line with the work and stance of the ETB, the publications did not discuss the children's recreational stays in France. Instead, Ackerman and Tchertkoff used the twentieth anniversary to increase awareness of the health and living conditions of the people in the most affected areas of Eastern Europe. In addition, Tchertkoff's book *Le crime de Tchernobyl: Le goulag nucléaire*[136] (The Chernobyl crime: the nuclear Gulag) described the wider nuclear politics at stake in the cover-up of the 'true' health impact of Chernobyl. From Tchertkoff's point of view, CORE and ETHOS essentially barred any critique of the evaluation policies and in particular hindered any challenges to the concept of radiophobia. The author asserted that it was therefore crucial to support people like Nesterenko and Bandazhevsky who opposed these policies. This stance was entirely within the nuclear political sphere, a setting that differed radically from how the British solidarity movement exponents communicated their work in 2006.

As mentioned, very few British publications on Chernobyl were released on the accident's twentieth anniversary. This reflects the minor role that Chernobyl has played in the British nuclear discourse. However, the few works published in 2006 are a good representation of the main actors in the British Chernobyl debate since the mid-1990s. Aside from Chris Busby, a central figure in the British solidarity movement published work on Chernobyl: Linda Walker, the head and founder of CCP (UK). Her book *Living with Chernobyl: Ira's Story*[137] appeared in the book series 'Real Life Stories' alongside other titles like *Refugee Camps, Street Children* and *Aids Orphans*. The book presented the life story of Ira, a 'Chernobyl child' living in a Belarusian orphanage. In terms of statements on the wider implications of Chernobyl, the book's very last page quoted scientists from the Ukrainian Ministry of Health to the effect that 'the Chernobyl radiation accident is undoubtedly the greatest environmental catastrophe in the history of mankind'.[138] In the final sentence, Linda Walker added: 'The world must never forget the children of Chernobyl, the generation yet to come and the terrible events of Saturday, 26 April, 1986'.[139] Apart from this conclusion, the book spoke almost exclusively about Ira and her daily life and friends in the orphanage. The accident was barely even mentioned, let alone the (nuclear) political issues relating to its impact. The author did not question the fact that the accident was rarely discussed with the orphans; instead she presented this setting as a given:

> Millions of people from Belarus, the Ukraine and parts of Russia, live with the consequences of the Chernobyl disaster every day of their lives. At places such as Zhuravichi and Rechitsa [names of orphanages] however, the accident and its effects are rarely discussed. The staff feel that the children have enough to worry about coping with their disabilities and coming to terms with what difficult future is ahead of them.[140]

Regarding statements on the health impact of Chernobyl, Walker's account was very reserved and did not provide any numbers at all, not even estimated ranges. Instead, she said: 'The number of casualties therefore remains controversial, but experts now agree that the Chernobyl disaster caused many cancers and, in particular, thyroid cancer. The full medical impact of Chernobyl will not be known until at least 2012'.[141]

It is not possible to say whether such opaque formulations regarding the health impact or absence of open allusions to nuclear politics – specifically regarding the contestation of the official evaluations of the most affected areas – were conditions dictated by the publishing house or were the author's decision. In any case, the exclusion of these topics has resulted in an almost exclusive focus on the individual. As a result, Ira's daily life was presented as something detached from Chernobyl. Of course, at one point there had been a connection, but this was an event in the past. In this regard, the brief synopsis on the back cover is interesting: 'Ira was born with severe disabilities two years after the Chernobyl disaster. She was given away to a home for abandoned babies and spent the next 11 years lying in a cot with nothing to do. But Ira has a huge strength of character – all she needed was to be given a chance'.[142] This statement was a declaration of the narrative's perspective; it effectively severed any connections her story had with nuclear issues. By emphasizing the personal character of children – this also occurred with Igor – it was possible to insinuate that the real scandal in their lives was their living conditions in Eastern European orphanages, which did not allow them to develop their capabilities and talents. When we place this type of narrative alongside that given in Tchertkoff's *Le crime de Tchernobyl*, it becomes even more obvious how apolitical (regarding nuclear politics) the publications of British solidarity movement exponents are compared to their French counterparts.

The book on Ira underpins my argument that the British solidarity movement groups' focus on the personal stories of the 'Chernobyl children' actually separated the topic of Chernobyl from the nuclear discourse. This detachment clearly facilitated these groups with their fundraising activities and explains why the solidarity movement became so strong in the UK, mainly regarding the number of children coming on holidays. At the same time, this detachment from nuclear issues promoted a 'denuclearization'[143] of the topic of Chernobyl in the British discourse. Additionally, this detachment counteracted the perception of personal affectedness.

Conclusion

In France, various actors in the lively French debate on Chernobyl used its twentieth anniversary to convey their statements on the impact of the accident to a broader public. This timing presented a particularly convenient opportunity for anti-nuclear activists to foster a debate on the use of nuclear energy in general. From their standpoint, this had become especially necessary due to French and international attempts to expand and advance the nuclear enterprise. French public authorities, on the other hand, used the occasion to lobby for their evaluation of the accident's impact. The French efforts to 'green the atom' – to present it as a sustainable environmentally friendly form of energy – and the new EPR in Flamanville, are the decisive historical context elements of the selective French Chernobyl debate in 2006. In Britain, the Chernobyl debate never reached anywhere near the same level of importance as in France. From the mid-1990s onwards, Chernobyl had basically vanished from the British nuclear discourse and only lived on in relation to charity activities. However, when the British government paved the way for a return to nuclear via new build in 2006, the topic of health effects from the fallout resurfaced in several newspaper articles and in the work of anti-nuclear activist Chris Busby, who used the anniversary to challenge national nuclear policies like French anti-nuclear activists had done. In France, a broad coalition of actors challenged the official Chernobyl narratives, whereas in Britain, only a handful of people publicly engaged in the Chernobyl debate in 2006. At the same time, the solidarity movement groups' activities dominated British media reporting. In contrast, the French solidarity movement groups were hardly reported on by the French media, which focussed on the health effects of Chernobyl: in Eastern Europe as well as in France.

The different focus in the two national contexts in 2006 reflects how Chernobyl has been debated differently in France and in Britain. In France, Chernobyl has become an anti-nuclear and anti-*nucléocratie* argument. Conforming with this trend, the twentieth anniversary generated a lot of attention, and anti-nuclear activists turned Chernobyl into a case against the government's EPR policies. At the same time, the anniversary was used to challenge the official assessment of the accident's impact – to which the public authorities responded by backing this official assessment. In Britain, however, Chernobyl did not embody a specific meaning with regard to national nuclear politics; this is also why British public authorities saw no need to communicate anything on Chernobyl in 2006. From the mid-1990s onwards, Chernobyl has been perceived as an event that only impacted people – and mostly children – in Eastern Europe. Because of this perception, Chernobyl could not be mobilized as an argument against British nuclear new build in 2006. The fact that the argument just simply did not work in the British context is well reflected in the feedback Busby received: insisting that there had been discernable Chernobyl health effects in Britain only further excluded him from the British nuclear debate because his views were widely considered as absurd.

Given this setting, it is interesting to examine the British debate on nuclear new build in the years following 2006: the anti-nuclear argument dominating the debate has not focussed on potential health effects but revolved around the issue of money. People have wondered whether this investment of tax money to subsidize a guaranteed price per kW/h for operators will pay off in the end. In France, the nuclear issue has been taken back to the streets and debated through such popular instruments as manifestations, rallies, blocking castor transports and the occupation of nuclear power plants. In Britain, however, the arguments against nuclear new build have been put forward by economists and political scientists in editorial articles and academic papers – that is, with the noteworthy exception of Scotland. Since its election in 2007, the Scottish National Party's government has opposed Westminster's new build plans and categorically refused to allow any new build projects on Scottish soil. The Scottish government was backed in its position by the Scottish Parliament, which re-formed following a referendum in 1997.[144] Thus, when discussing the 'British' position in the following paragraphs, I refer to a Britain that predominantly represents the government in London's stance.

In France, Chernobyl can be well integrated into the anti-nuclear discourse because its memory inherently implies criticism of the French *nucléocratie*, and – due to the strongly perceived personal affectedness – there is a large audience for health-related nuclear arguments. In Britain, however, the argument over the potential health effects of low-level radiation exposure and an oppressive *Atomstaat* (the nuclear state, in reference to Robert Jungk's homonymous book) – both mounted in relation to the civil nuclear enterprise – is voiced by a very small group of people. Thus, Chernobyl as an anti-nuclear contention is quite an outsider argument. In addition, the already very small group of anti-nuclear power plant activists in Britain has shrunk even more since the early 2000s. Climate change came on the agenda as the new environmental buzzword. Claiming that the climate challenge necessitated a reconsideration of issues related to anti-nuclear positions, George Monbiot left the British anti-nuclear movement to become one of Britain's most prominent advocates of nuclear new build. A strong believer in the narrative of the 'green atom', Monbiot came to consider the anti-nuclear movement as the enemy. This shift by environmentalists from the anti- to the pro-nuclear side has also been verified in other countries than Britain. But it seems that in Britain, the pro-nuclear environmentalists came to dominate the nuclear discourse. This might explain why, regardless of the internationally supported re-emergence of the Chernobyl debate, the twentieth anniversary could not spark a broader debate on British Chernobyl personal affectedness or mobilize a larger group of people to campaign against the renaissance of the British nuclear enterprise. Britain seems to lack a driving force capable of turning events such as the British government's attempt to manipulate media reporting on Fukushima – as Rob Edwards revealed in the *Guardian* in July 2011 – into a more widely backed argument against national nuclear policies. In other countries, this driving force consists of environmental groups. In Britain, however, this important actor has partly withdrawn from the anti-nuclear side and has never tried to gain a prominent role in the British Chernobyl debate.

This is not to say anti-nuclear voices have been completely silent in Britain in recent years, but compared to many Western European countries such as France, Germany or Italy, nuclear energy remained a marginal topic within the political discourse. Moreover, Chernobyl never became a central argument in the British anti-nuclear power discourse, which always focussed on the Sellafield-Windscale complex. Interestingly, this continuity has also shaped the way meaning is assigned to Chernobyl. When Chernobyl was discursively connected to British nuclear plants, it adhered to the argument: Sellafield is x-times more dangerous than the Chernobyl fallout.[145] This scheme can also be found in France, mainly applied to the scenario of a plane crashing into the La Hague site: which would cause x-times the damage of Chernobyl. But most frequently – and this is a very interesting difference between the British and the French nuclear discourses – the French comparison is the other way around: Chernobyl poses x-times a bigger risk than situation y. This reversed line of reasoning is illustrated by the IRSN's panel study on risk perception. In 2006, 50 per cent of the interviewees perceived the nuclear fallout caused by Chernobyl to be an 'elevated risk'. This meant, at the same time, that this fall-out was perceived to be riskier than household accidents, genetically modified foodstuffs, or nuclear power plants in general.[146] Thus, Chernobyl and its health effects on France have become topics in themselves there. In Britain, however, Chernobyl and its health effects on Britain have not even become central topics in the nuclear discourse. This discourse has also further declined since the early 2000s as the climate change context has led to a 'denuclearization' of the British nuclear discourse. Consequently, many main aspects of the debate about nuclear new build are not framed as nuclear issues but are instead linked to economics and global warming.

Notes

1. The Chernobyl Forum, *Chernobyl's Legacy*, 14ff.
2. Ibid., 19ff.
3. Ibid., 36ff.
4. Ibid., 37.
5. The WHO report – which lay at the basis of the Chernobyl Forum report – also stressed the psychological effects of the accident. In its 2006 publication, the WHO's Chernobyl Forum Expert Group 'Health' stated: 'the mental health impact of Chernobyl is the largest public health problem caused by the accident to date'. (WHO, *Health Effects of the Chernobyl Accident*, 95). This expert group concluded: 'The accident has had a serious impact on mental health and well-being in the general population. Importantly, however, it appears that this impact is demonstrable mainly at a sub-clinical level. Although the empirical studies do not support the view that the public anxiety bears a resemblance to clinical psychiatric disorders, such as phobia or psychosis, the disaster did have a psychological effect that is not limited to mental health outcomes. It also has ramifications for other areas of subjective health and health-related behaviour, especially reproductive health and medical service utilization, and the level of trust in authorities. Further, it may influence people's willingness to adopt safety guidelines issued by the authorities' (ibid., 96).
6. The Chernobyl Forum, *Chernobyl's Legacy*, 8.
7. For CRIIRAD's commitment to support Bandazhevsky after his release from prison, see: CRIIRAD, *Trait d'Union*.
8. For these legal battles, see: Kalmbach, *Tschernobyl und Frankreich*, 114ff.
9. Bandajevskaya, 'Comment on a réduit au silence'; David-Jougneau, 'Semmelweis, Bandajevsky'.
10. Tchertkoff, *Le crime de Tchernobyl*, 349ff.
11. Lhomme, *L'insécurité nucléaire*, 194.
12. Desbordes, *Atomic Park*, 279ff. For an analysis of the book, see: Kalmbach, *Tschernobyl und Frankreich*, 161.

13. Montellier, *Tchernobyl mon amour*, 29.
14. Ibid.
15. Bandazhevsky, *La philosophie de ma vie.*
16. Ibid., 15.
17. Ibid., the book's back cover.
18. Bandazhevsky et al., *Tchernobyl 25 ans après.*
19. Bandazhevsky and Dubovaya, *Les conséquences de Tchernobyl sur la natalité.*
20. Bandazhevsky and Bandajevskaya, *Les conséquences de Tchernobyl sur la santé.*
21. The ETB also supported the Bandazhevskys' work, for instance by collecting money from ETB members for Galina and her daughters, see: Enfants de Tchernobyl Bélarus, 'Compte rendu de l'Assemblée Générale Ordinaire'.
22. Nesterenko, 'L'Europe aurait pu devenir inhabitable'.
23. Alexey Yablokov (together with Iryna Labunska and Ivan Blokov) was also general editor of the report commissioned by Greenpeace, *The Chernobyl Catastrophe.*
24. Yablokov and Nesterenko and Nesterenko, *Chernobyl: Consequences of the Catastrophe.*
25. See for a detailed analysis of the controversy around the Yablokov-Nesterenko report: Kalmbach, 'The Contested Truth', 252–7; Bauer and Kalmbach and Kasperski, 'From Pripyat to Paris'.
26. Yablokov and Nesterenko and Nesterenko, *Chernobyl: Consequences of the Catastrophe*, 50.
27. Ibid., 195.
28. Cotterill and Pearce and Parker, 'Thyroid Cancer in Children and Young Adults'.
29. Ibid., 1020, abstract.
30. For the post-Fukushima dispute between Busby and Monbiot, see: Kalmbach, 'The Contested Truth', 254–6.
31 See: Kalmbach, 'Radiation and Borders'.
32. This statement stems from my analysis of the search results in newspaper database Newsbank using the search key 'Chernobyl' for the period between 15 and 30 April 2006.
33. I conducted a similar search as described in the note above for the tenth anniversary, between 15 and 30 April 1996. However, the entries do not reflect the actual number of articles on Chernobyl, because many from 1996 were not yet in the Newsbank database.
34. For a detailed analysis of French newspaper coverage of the twentieth anniversary, see: Kalmbach, *Tschernobyl und Frankreich*, 168–71.
35. Meikle, 'Sheep Farms under Curbs'.
36. This number dropped further in the following years. According to the *Guardian* (Hickman, 'What Is Safe to Eat?'), in 2009, 'there were still 355 farms in Wales (in and around Snowdonia), 9 in England and 7 in Scotland. All the farms in Northern Ireland were "derestricted" in 2000'.
37. Meikle, 'Sheep Farms under Curbs'.
38. Jowit, 'Chernobyl's Generations of Suffering'.
39. Page, 'Chernobyl Hero'.
40. *Western Daily Press*, 'Nature's Cure for Nuclear Fallout'.
41. *The Sun*, 'Life Line to Radiation Kids'; *Hull Daily Mail*, 'We Won't Forget Chernobyl Victims'.

42. Bolouri, 'Nuclear Family of Love'.
43. *The Independent on Sunday*, 'Still Causing Cancer in British Children'.
44. Forgrave, 'Wales: Did Chernobyl Disaster Cause My Cancer?'.
45. Simons, 'Ill Winds Carried Chernobyl Poison'.
46. Forgrave, 'Wales: Did Chernobyl Disaster Cause My Cancer?'.
47. Ibid.
48. *Wales on Sunday*, 'Remembering Chernobyl'.
49. A possible connection to Sellafield-Windscale arose even in the debate on increased cancer rates on Benbecula. *The Herald Scotland* of 1 April 1996 (Buie, 'Cancer Island Inquiry') referred to an early 1990s study on the Western Isle North Uist, revealing that 'some islanders had five times more radioactivity than people elsewhere in Scotland. At the time, the excess levels of caesium found in the islanders was attributed to discharges from the Sellafield nuclear processing plant entering the food chain in the Western Isles'. However, other 'possible links with the firing of rockets, possibly tipped with radioactive material, from the military range at Benbecula and the presence of a powerful early warning radar system' were also discussed.
50. Highfield, 'Windscale Fallout'.
51. This paragraph is a summary of the corresponding chapter in: Kalmbach, *Tschernobyl und Frankreich*, 134–7.
52. CEA, 'Accident de Tchernobyl'.
53. Bigot, *Point sur les conséquences sanitaires.*
54. Ibid., 3.
55. Ibid., 6.
56. Ibid., 14.
57. Ibid.
58. Ibid., 18ff.
59. Ibid., 18.
60. Ibid., 13.
61. Ibid., 23ff.
62. The following account is adapted from: Kalmbach, *Tschernobyl und Frankreich*, 138–40.
63. IRSN, 'Les leçons de Tchernobyl' (the following notes refer to the dossier's sub-chapters).
64. IRSN, 'Les leçons de Tchernobyl – Evaluer les conséquences et protéger les personnes'.
65. IRSN, 'Les leçons de Tchernobyl – Exposition des personnes en France'.
66. IRSN, 'Les leçons de Tchernobyl – Mieux agir en territoire contaminé'.
67. Renaud and Champion and Brenot, *Les retombées radioactives.*
68. Ibid., ix.
69. Ibid., 146.
70. Ibid., 149.
71. Ibid., 153.
72. This is adapted from: Kalmbach, *Tschernobyl und Frankreich*, 140–41.
73. InVS, *Surveillance sanitaire en France.*
74. The results of the study on Corsica were published in 2012: InVS, *Estimation de l'incidence du cancer de la thyroïde en Corse* (Estimation of thyroid cancer incidence in Corsica). The InVS website formerly provided an English summary of the study: 'In France, Corsica appears to be one of the most exposed regions to the fallout from the

Chernobyl accident. Taking into account the scientific knowledge at that time, it was decided to focus studies on thyroid cancers. A study was carried out in order to estimate thyroid cancer incidence in Corsica for the periods 1998–2001 and 2002–2006. The study identified incident thyroid cancer cases between 1998 and 2006 among residents in Corsica. Data were collected using information from the hospitals (PMSI) and the local health insurance funds (ALD). Cases were validated through medical records before inclusion in the study. Over the period of study, 342 cases of thyroid cancer, rather women and relatively young patients, were identified in Corsica. Incidence rate of the thyroid cancer was high, but stable among men, and with a slight increase among women, particularly between 2002 and 2006. However, incidence rate and clinical characteristics of thyroid cancer in Corsica are not exceptional and are similar to those in other French districts'. On: http://www.invs.sante.fr/Publications-et-outils/Rapports-et-syntheses/Maladies-chroniques-et-traumatismes/2012/Estimation-de-l-incidence-du-cancer-de-la-thyroide-en-Corse (last accessed 15 August 2018).
75. InVS, *Surveillance sanitaire en France*, 65.
76. Ibid.
77. Ibid., 66.
78. Ibid., 4.
79. Ibid., 3.
80. I do not know if the HPA had information on Chernobyl on its website in 2006. The account on its 2013 website was brief and 'Chernobyl' was not even a key word in the topic list of several hundred entries. A quarter of the text explained 'The Impact in the UK': 'Chernobyl fallout was first detected in northern Poland on Monday, 28 April, 1986 and soon afterwards across the Baltic Sea in Sweden. The plume then travelled southwards over Germany and up through France to the UK, arriving in southeast England on the morning of Friday, 2 May. The plume had travelled over 1000 miles before reaching the UK and so was considerably dispersed. It was nevertheless very detectable with suitable instruments and there was some widespread "dry" deposition (i.e. not caused by rainfall). Rainfall in North Wales, NW England and Scotland deposited more fission products on the ground via "wet" deposition. Iodine-131 was measured in air at about 1 or 2 becquerels per cubic metre and could be detected in milk products in these and other regions, but not at levels that would have triggered a ban on consumption. Caesium-137 could also be detected in soil samples and this still has an impact in parts of the UK. In most areas the caesium-137 was washed into the ground where it binds strongly with clay and other soils and is therefore removed from the food chain. However in certain parts of the UK the soil and grasses combine to recycle the caesium-137 and it can then enter the food chain. This occurs in some hill farm areas and has required restrictions on produce from sheep farms in these areas. Measurements were carried out to see how much radiation people had absorbed during, and after, the passage of the plume. In southern UK, people had levels of about 300 becquerels of caesium-137 in their bodies in 1986 which dropped to 50 becquerels in 1989 and then were back to pre-incident levels in 1990. For comparison, adults have about 4,000 becquerels of naturally occurring potassium-40 in their bodies. Potassium is essential for life and potassium-40 emits electrons and a gamma ray with a very similar energy to the gamma ray from caesium-137'. The text was formerly available at: http://www.hpa.org.uk/Topics/Radiation/Under

standingRadiation/UnderstandingRadiationTopics/RadiationIncidents/Chernobyl/ (last accessed 15 June 2013).
81. Health Protection Agency, 'The Consequences of the Chernobyl Nuclear Accident'.
82. Until 2008, Cardis was Head of the Radiation Group at the International Agency for Research on Cancer, the cancer agency of the WHO.
83. O'Riordan, *Radiation Protection.*
84. Ibid., 119.
85. Ibid., 120.
86. See: Mühlenhöver, *L'environnement en politique étrangère.*
87. Lauvergeon and Jamard, *La troisième révolution énergétique.*
88. Ibid., 132.
89. Ibid.
90. Ibid., 132ff.
91. Ibid., 133.
92. Ibid., 131.
93. Greenpeace France, *20 ans Tchernobyl.*
94. Until 2011, information on this CD as well as a streaming function for all titles were available on Greenpeace's website: http://www.greenpeace.org/france/news/CD-20ans-tchernobyl. General information on the CD is available at: http://musique.fnac.com/a1840356/Variete-francaise-20-ans-Tchernobyl-CD-album#ficheDt (last accessed 15 February 2020).
95. A website listing all of Greenpeace's Chernobyl-related events in Paris was available until 2011 at: http://www.greenpeace.org/france/news/expo-tchernobyl.
96. Until 2011, the website was available at: http://www.sortirdunucleaire.org/index.php?menu=sinformer&sousmenu=themas&souss ousmenu=tcherno3&page=index.
97. Stéphane Lhomme, *L'insécurité nucléaire.*
98. For a detailed analysis of Lhomme's book, see: Kalmbach, *Tschernobyl und Frankreich,* 151–5.
99. Lhomme, *L'insécurité nucléaire,* 183.
100. The ETHOS project and CORE programme aimed to solve the high stress levels of inhabitants dealing with new, unknown and invisible radioactive hazards on a daily basis, and their mistrust of the authorities and experts. For an analysis of the implications, see: Bauer and Kalmbach and Kasperski, 'From Pripyat to Paris'.
101. Lhomme, *L'insécurité nucléaire,* 185.
102. Ibid., 196ff.
103. Ibid., 205.
104. This paragraph is adapted from: Kalmbach, *Tschernobyl und Frankreich,* 150.
105. CRIIRAD, *Tchernobyl – 20 ans après.*
106. Ibid., 2.
107. This debate was announced on CRIIRAD's website, however, as far as I know, it never materialized.
108. These books are all discussed in: Kalmbach, *Tschernobyl und Frankreich,* 144–8.
109. Ackerman and Grandazzi and Lemarchand, *Les silences de Tchernobyl.*
110. Ackerman, *Tchernobyl, retour sur un désastre.*
111. Kostine, *Tchernobyl: confessions d'un reporter* (Chernobyl: confessions of a reporter). Versions of the original book in French appeared in the same year in English, German,

Spanish and Italian. The English edition was published by American publisher Umbrage Editions. The book was the result of a collaboration between Igor Kostin, film producer and director Thomas Johnson, and Galia Ackerman. Johnson also produced a film about Igor Kostin and his famous photographs of Chernobyl: *La bataille de Tchernobyl* (The battle of Chernobyl). Both the book and the film were widely received in France on the twentieth anniversary. Kostin's interpretation is the visual analogue to Alexievich's narratives: his images show the effects of an event that has brought unimaginable suffering to humanity, the full extent of which is not yet known, as the mutagenic effects of radiation will only be fully manifest in future generations. For the twentieth anniversary, the *Guardian* compiled an online exhibition of Kostin's most widespread photos: *The Guardian*, 'Chernobyl Nuclear Disaster – in Pictures'. The introductory text to this exhibition states: 'His images of a deformed boy even led to adoption of the "Chernobyl Child" in UK'. In fact, this boy was Igor Pavlovets.

112. Grandazzi, 'Die Zukunft erinnern'.
113. Lemarchand et al., *Destination Tchernobyl.*
114. Dupuy, *Pour un catastrophisme éclairé.*
115. Dupuy, *Retour de Tchernobyl.*
116. Fairlie and Sumner, *The Other Report on Chernobyl.*
117. See the EDF's press release of 23 May 2007 which was formerly available at: http://medias.edf.com/communiques-de-presse/tous-les-communiques-de-presse/communiques-2007/edf-apporte-son-experience-et-ses-competences-nucleaires-aux-britanniques-40311.html (last accessed 15 June 2013).
118. The Sussex Energy Group was one of the fiercest critics of the British government's plans, see: Sussex Energy Group, 'Response to Government's "The Future of Nuclear Power" Consultation'.
119. Greenpeace UK, 'Hinkley Strike Price Briefing'.
120. Busby and Yablokov, *Chernobyl 20 Years on.*
121. Ibid., 3.
122. Ibid.
123. Ibid., 2.
124. Ibid., 247.
125. Ibid., 136.
126. These cluster studies investigating the health situation of children living near nuclear sites have been conducted in various countries and were mostly initiated by anti-nuclear groups. They often found an 'abnormal level' of specific illnesses in children; all these studies and their applied methodologies are highly contested, especially by pro-nuclear scientists.
127. Busby and Yablokov, *Chernobyl 20 years on*, 142.
128. As the analysis of French publications released for the twentieth anniversary forms a central part of my book *Tschernobyl und Frankreich*, the account given here is brief.
129. Jacquemin-Raffestin, *Tchernobyl 20 ans après.*
130. For a detailed analysis of Jacquemin's 2006 publications, see: Kalmbach, *Tschernobyl und Frankreich*, 156–8.
131. Montellier, *Tchernobyl mon amour.* For an analysis, see: Kalmbach, *Tschernobyl und Frankreich*, 165–6.

132. Micoulaut, *Tchernobyl. L'histoire d'une désinformation.* For an analysis, see: Kalmbach, *Tschernobyl und Frankreich*, 162–3.
133. Kostine, *Tchernobyl: confessions d'un reporter.* For an analysis, see: Kalmbach, *Tschernobyl und Frankreich*, 159–61.
134. Bandazhevsky, *La philosophie de ma vie.*
135. Anders, *La menace nucléaire.*
136. Tchertkoff, *Le crime de Tchernobyl.*
137. Walker, *Living with Chernobyl.*
138. Ibid., 45.
139. Ibid.
140. Ibid., 30.
141. Ibid., 44.
142. Ibid., the book's back cover.
143. The 'denuclearization' of Chernobyl's consequences is an essential part of Chernobyl politics in Eastern Europe. See: Bauer and Kalmbach and Kasperski, 'From Pripyat to Paris'. The term 'denuclearization' refers to Gabrielle Hecht's concept of 'nuclearity'.
144. For the Parliament's decision, see: *The Scotsman*, 'MSPs Vote No to Nuclear Stations'.
145. The comparison to Sellafield is also central to the Irish Chernobyl debate. The Irish and British Chernobyl debates are tightly interwoven, not least via individuals like Adi Roche and publications on Chernobyl circulating across the Irish Sea.
146. IRSN, *Baromètre IRSN 2006.*

Conclusion

The comparison between the French and British Chernobyl debates clearly shows that 'the degree to which a country is affected by the hazard'[1] is, with regard to Chernobyl, not proportional to the radioactive fallout that the country received. Risk exposure cannot be understood in terms of the absolute number of radionuclides deposited on a certain region or taken up by an individual. This approach has proven useful for explaining the short-term changes in public opinion polls after Chernobyl[2] (given that it was taken for granted that the intensity of the fallout was communicated correctly). However, this approach cannot explain why in one country, France, an intense debate on the health impact of Chernobyl was re-sparked in 1996, whereas in Britain, a country with a comparable deposition of radionuclides (as far as may be discerned from the published figures), such a public debate never rose to prominence. Therefore, the question is not: how much fallout was deposited on a certain country or region? but rather: how has this fallout been perceived, and in what context? Thus, the interpretation of Chernobyl is not so much the result of a direct physical impact, but more a crystallization of existing sets of beliefs. They determine how the accident has been perceived and narrated, or not been considered at all.

By analysing the various actors' narratives in the national Chernobyl debates, I was able to shed light on how Chernobyl has been described, perceived and interpreted in France and Britain. My analysis has focussed on the narrative elements of radiophobia and apocalypse, personal affectedness and anti-East European/anti-Soviet stereotypes. This latter aspect played a more important role in the comparison between different narratives within a single national context than when comparing the two national case studies. Therefore, this conclusion focuses on my analysis of radiophobia versus apocalypse and personal affectedness.

The narratives that developed around the health effects of Chernobyl can be situated on a spectrum between two extremes. At one extreme lies the explanation pattern of radiophobia and at the other, an apocalyptic narrative that describes Chernobyl as a type of allegory for the end of the world.[3] The protagonists in the Chernobyl debate normally took a clear stance on health effects caused by the fallout and aligned their statements closer to one of the two extremes. Therefore, the initial questions when examining the various sources for this category of comparison were: who said what? What interpretations were proffered by national public authorities and actors in the national nuclear industries? Did a specific counter narrative emerge in response to the official experts' evaluations, and if so, who paid attention to it; who believed it?

Secondly, investigating the narrative element of personal affectedness made it clear to me why Chernobyl became such an important topic and reference point in the French nuclear debate, while in Britain it almost sank into total oblivion before partly resurfacing in 2006. The detailed analysis of British Chernobyl literature that I conducted for this study should not give the reader the impression that Chernobyl was a prominent topic in the British public discourse.[4] A purely quantitative comparison between the French and British debates would only lead to the conclusion that there are barely grounds for a reasonable comparison, especially regarding the twentieth anniversary. A quantitative observation reveals very little about the debates themselves, and nothing at all about the reasons for this dramatic difference. However, conducting a qualitative comparison and looking more closely at the structure of the Chernobyl narratives in the two national contexts, revealed very interesting findings. To achieve this type of comparison, I had to analyse in detail the accounts of the various actors in the French and British Chernobyl debates. The British material is presented for the first time, whereas most of the French material has already been examined in my book *Tschernobyl und Frankreich*. In order to present the research clearly and concisely, I divided the development of the debates into three distinct time-slots: the accident and its direct aftermath, the trajectory of the debate from 1989 to 2005 and the debate on the accident's twentieth anniversary. These three stages had major markers: for both countries, the debate in the direct aftermath of the accident was very similar. However, in the following stage, the two trajectories split. The British debate fell off sharply in the mid-1990s, while the French debate began to expand and evolve. For both countries, the mid-1990s were an important turning point: from this time on, the aspect of personal affectedness in Britain almost entirely disappeared from the Chernobyl debate, whereas in France, personal affectedness experienced a lively awakening from the mid-1990s. We should see this changing perception of personal affectedness in light of both countries' evolving nuclear policies. Finally, in the third stage, a closer look at the specific debate on the twentieth anniversary of the accident clarified to what extent national nuclear politics and policies and the perception of personal affectedness are interrelated in the Chernobyl debates.

National nuclear politics and policies were not impervious to external influences, and so while these politics and policies shaped the respective Chernobyl debates, they were also in turn shaped and influenced by this debate. Many more aspects influenced the different trajectories of the British and

French debate. For this reason, I systematically located these debates within their historical context and referred to the national specificities that influenced their development and tenor. The aspects I focussed on were: the formation, role and status of nuclear experts and 'counter experts'; the (changes to) national nuclear politics, policies and polities as well as their pro-nuclear versus anti-nuclear orientations; the anti-nuclear movement's structure, political role and protest culture; the (issues with) national nuclear plants; and the importance of charitable organizations. In the following paragraphs, I will summarize the results of this contextualization. The conclusions drawn from comparing the French and the British case are manifold. Most importantly, no single factor can explain why the debate on the impact of Chernobyl developed differently from one country to the other. A broad range of factors came into play – factors that reciprocally influenced each other.

At first glance, the debates on Chernobyl and their ramifications might seem likely to proceed similarly in France and Britain. Each country was affected by a comparable degree of fallout in late April and early May 1986, and each had a highly developed nuclear sector. Both countries had an active anti-nuclear movement and had previously experienced accidents in their own national nuclear plants. Members of each government proclaimed that the fallout from Chernobyl would have no relevant impact on their national territory – in both countries they were proven wrong. The French and British governments, radiation protection agencies, and various actors in the nuclear industry argued that an accident such as Chernobyl could not happen in their own country. Later, in both countries, local doctors – in the French case in Corsica and in the British case on Benbecula – suggested a possible connection between increases in the cancer rates they observed in their daily work and a possible health impact from the fallout. Finally, in both countries, individual actors – Jean-Michel Jacquemin in France, Chris Busby in Britain – picked up this topic and spent a great deal of time seeking out national Chernobyl victims.

Despite these many similarities, the relative impact of the Chernobyl debates in France and Britain could not have been more different. In France, Chernobyl became a national reference point for criticizing the central government and the country's political and scientific elite system. Conversely, in Britain, the Chernobyl solidarity movement worried about maintaining support for its work, as public memory of the accident faded. In France, Chernobyl became a contemporary representation of threats to everyday life, while in Britain, it was relegated to the past; considered an accident that happened far away and long ago. Few people remember the extensive restrictions on British sheep farms in the aftermath of Chernobyl because the animals were too radioactive to be sent to market.

To explain this process – the formation of a national *lieu de mémoire* linked with the political scandal of the *affaire Tchernobyl* in France and the non-existence of this common point of reference in Britain – we must consider various aspects. The first and most important aspect seems to be how national official experts and their evaluation of the fallout's impact were perceived by the public. As Brian Wynne argued, the interaction between experts and the public is highly context-bound:

> Though they pervade all processes of 'understanding', trust and credibility are contingent variables which depend upon evolving relationships and identities. … The fundamental interaction between scientific expertise and lay-publics is cultural, in that scientific knowledge embodies social and cultural prescriptions in its very structure. The problems of public uptake of science therefore lie in the institutional forms of science and of its incorporation into policy and administration.[5]

In France, a rather large proportion of the public adopted a critical stance towards the government and its nuclear experts, after their claims that the French people had not been exposed to any risk. The French public's distrust was fuelled by a more general criticism of the French elite system, specifically the exclusive positions graduates of the Grandes écoles hold in the upper echelons of the state administration and their supposed detachment from the rest of society. Long before Chernobyl, the French *nucléocratie* had been accused of being a closed circle of fortunate and privileged individuals who cared about their own career above all else. From the start, this general questioning of key structures in the French nuclear sector was drawn into the debate on the health effects of Chernobyl. The fact that state experts actively engaged in the debate on French personal affectedness increased a broader audience's attention to this topic, and at the same time continually instigated counter statements from critics. Confronted with a French government promoting nuclear power at the global level as a green solution to climate change, the French anti-nuclear side transformed the debate on the health impact of Chernobyl into a proxy war over legitimizing the nuclear enterprise.

Although the reactions of the British government and its nuclear experts in 1986 were similar to those in France, several important differences in the contextual setting account for the different Chernobyl debate in Britain. One decisive aspect is how nuclear experts and their role in British society were perceived. For example, with universities like Oxford and Cambridge, the UK has a longstanding tradition of a distinct scientific and political elite. However, the experts in the civil nuclear industry – be they engineers or physicists – were never assigned the label of a group illegitimately detached from average people. They fulfilled a significant role in society and, because of their education, were considered an authority in scientific evaluations. This fact, however, was never perceived to be as much of a problem as in France. Therefore, the criticism levelled by actors challenging the official evaluations did not find in Chernobyl the fertile ground to disseminate their argument to a greater and more receptive audience. As I was able to demonstrate, some critical voices that questioned the narratives presented by state experts did exist in Britain. But they were few and far between. When the government passed the decision in the mid-1990s to cease nuclear new build projects, these voices lost even more ground. At this point, unlike their French counterparts, British anti-nuclear power activists did not need to turn the debate on Chernobyl health effects into a battle over legitimizing the nuclear enterprise, because their (national) fight seemed to be won already. A 'discursive gap' formed that would be filled by the solidarity movement, and thus Chernobyl as a *lieu de mémoire* lost its anti-nuclear connotations in Britain. Consequently, Chernobyl became an accident that happened far away and long ago, an exciting event that could entertain British children and teenagers in non-fiction books without causing them to fear their own environment. The slow re-emergence of Chernobyl as a topic of public debate, when government support for nuclear power re-emerged in 2006, reveals how closely national debates on the impact of Chernobyl clung to national nuclear policies.

Based on the analysis of these elements thus far, expert culture in the civil nuclear industry could be identified as the main factor shaping the national Chernobyl debates in France and Britain. However, as my comparison of these two cases reveals, many other factors influenced the trajectories of these debates. The role of individual agency should not be underestimated. In France, particularly in terms of propagating personal affectedness, the key actors were Michèle Rivasi and her CRIIRAD colleagues, Jean-Michel Jacquemin and local doctors in Corsica. Without their involvement, the debate on the health effects in France would probably never have gained such importance. The dissemination of the apocalyptic narrative regarding the most affected areas of Eastern Europe – particularly spreading the word about Alexievich and Bandazhevsky's work – was also achieved by individuals. Without the activism and publications by individuals in the 'Caen-Group' and the CRIIRAD, these voices from Eastern Europe would have remained unknown in France. No such activism, nor indeed any comparable level of activism, exists in the British Chernobyl debate. Individual agency can also be found in the British case, for instance Chris Busby or the solidarity movement groups. But the solidarity groups neither lobbied for British personal affectedness nor for an apocalyptic narrative. Busby did, but within Britain, he remained an isolated campaigner. In France, however, the number and variety of Chernobyl activists are considerable. Moreover, these activists were able to publish their Chernobyl accounts in large publishing companies. Thanks to strongly perceived personal affectedness, there was a bigger market for Chernobyl literature in France than in Britain.

The analysis of the French and British Chernobyl debates reveals another interesting finding regarding individual agency. The existence of individual Chernobyl activists is not connected to or a result of the particular strength or degree of political institutionalization of the environmental movement. In both countries, the Green Party has played a marginal role in national politics. Both countries have close links between prominent individual Chernobyl activists and their respective national Green Party: Busby was a spokesperson for the Green Party and Michèle Rivasi has been an MEP since 2009. But in France, numerous political Chernobyl activists are not necessarily connected to the Green Party, and in Britain, Chernobyl activism is rather apolitical.

The environmental movements seem to have played a decisive role in their national Chernobyl debates in a different way. It is not so much the strength or extent of institutionalization that influenced the debates, but rather the way the environmental movement addressed the topic of climate change. Without generalizing George Monbiot's standpoint onto the British environmental movement, his determined fight against climate change deniers hints at an aspect of the British case that might differ from other countries. A 2009 opinion poll in Britain revealed that only 'some 41 per cent of those taking part in today's poll agreed that it has been established that climate change is largely due to human activity'.[6] This context might partly explain Monbiot's strong insistence to oppose climate change and to sensitize the public on this topic as a priority for environmental campaigners. At the same time, Monbiot has used climate change as an argument for attacking anti-nuclear campaigners because, in his opinion, their activities have hindered international efforts to combat global warming. And, as the publicized switching of sides by former anti-nuclear campaigners to the pro-nuclear camp clearly shows, Monbiot is not alone in his view.

Despite the diminishing anti-nuclear power movement due to the successful marketing of nuclear reactors as climate saviours, British anti-nuclear power protest has never been strong. British anti-nuclear campaigners have always focussed on the military applications of nuclear technology. And even before people like Monbiot changed sides, the small group of anti-nuclear power campaigners had already lost much of its influence after the moratorium on British new build in the mid-1990s. The rare instances of public anti-nuclear power criticism voiced after the mid-1990s have been directed at Sellafield, not Chernobyl. The Sellafield-Windscale complex has always been Britain's primary reference point with regard to personal affectedness and government cover-ups; and it has maintained this position after 1986. For this reason, many accounts on Chernobyl used Sellafield as the analytic framework or unit of comparison. The fact that one of the British regions most affected by the Chernobyl fallout, the Lake District, is also the region that houses the Sellafield site, has reinforced the discursive connection between these two nuclear threats. Furthermore, the restrictions on sheep farms in this region after Chernobyl caused the Windscale Fire to resurface in public memory. First, research in the Chernobyl fallout always included a reference to the Sellafield-Windscale complex, since the background radiation in this region was already higher due to contamination in 1957. Second, this setting resulted in the local perception that the countermeasures against the Chernobyl fallout were actually the delayed effects of the Windscale Fire: because the region already had a higher level of background radiation prior to the Chernobyl accident, dose limits were exceeded only because of the sum of the two fallouts.

British personal affectedness, therefore, was not so much the perceived victimization brought about by the events of 1986, but rather a personal affectedness linked to the (re)discovery of the permanent threat posed by nuclear installations in the UK, particularly Sellafield. We can find this interpretative pattern in very early critical accounts of Chernobyl as well as in later publications by Chris Busby. Attached to the account on 'The Chernobyl Fallout' in Mackay and Thompson's 1988 compilation *Something in the Wind* was a map showing the 'danger zones around nuclear reactors in Britain' instead of the Chernobyl fallout intensities in the UK.[7] Furthermore, Busby used the impact of Chernobyl in Britain as an argument to underpin his statements on the devastating health effects of British nuclear power plants. In France, however, Chernobyl was treated as a distinct and independent event. Although the Chernobyl debate, through the correlated criticism of the *nucléocratie*, was discursively closely connected to the French nuclear enterprise, the debate on the fallout's health impact did not lead to rediscovering for example the 1969 accident in Saint-Laurent. Moreover, as revealed by the 2006 IRSN opinion poll, the Chernobyl fallout in France was perceived to be a bigger threat than national nuclear installations. Thus, we are confronted with a situation that demonstrates opposite traits to Britain. Indeed, on the twentieth anniversary of the accident, the Chernobyl debate, within the context of the protests to the EPR project in Flamanville, referred to the threats from the French civil nuclear enterprise. But in 2006, public discourse still treated the health effects of Chernobyl in France as a separate topic, detached from health effects that might have been caused by the national nuclear fleet. This discrepancy points to a profound difference in the respective national attributions of nuclear risk. A simplified list ranking the risk attributions (from highest to lowest) in Britain would look something like this:

1. nuclear weapons
2. reprocessing plants
3. nuclear power plants
4. Chernobyl fallout.

In France, however, this ranking would be reversed, with Chernobyl fallout in first place and nuclear weapons in last.

In this regard, the argument on the 'externalization of fears' that Melanie Arndt applied to explain the reasons underlying West German Chernobyl solidarity activities can also be applied to the British case: inviting 'Chernobyl children' for holidays in the UK has helped the host families to externalize their worries about the nuclear installations close by on national territory. In France, however, the 'foreign' radiation and its impact on the country have formed the focal point of the public Chernobyl debate and many people have perceived the health risks of the fallout to be more dangerous than the presence of nuclear power plants just a few miles down the road. This risk attribution may be a long-term effect of the profound pride in national nuclear technical capability, which Gabrielle Hecht described for the French post-war period. The fact that employing foreign contract workers in French nuclear plants – the so-called 'nuclear nomads' who do the most radioactive dirty work – has long been a focus of criticism against French nuclear policies[8] might further underpin this hypothesis. Undoubtedly, the British post-war era experienced similar levels of pride in the national nuclear technical capability, immortalized by Queen Elizabeth II's inauguration of Calder Hall Power Station. But this came to an end with the privatization of the British nuclear industry and the sale of the British nuclear power plants to a French company, the EDF. In France, however, the government and EDF's lasting commitment to the strategy of *tout nucléaire*, the successful creation of Areva, the development of the EPR and the proclaimed 'nuclear renaissance' discursively (within the context of the climate change debate) transformed nuclear plants; from the grey monolithic projections of planning fantasies and the strong belief in technology-driven steady progress of the 1960s, they became the environmentally benign steam generators of the future. This setting created ideal conditions for the pride in national nuclear technical capability to thrive, and attribute nuclear risk to foreign entities, like the Chernobyl fallout. To link this attribution of nuclear risk to the almost total lack of focus on holidays in France for 'Chernobyl children' in the public Chernobyl debate, I present the following hypothesis: one of the reasons recreational stays for 'Chernobyl children' in France are not more diffuse, may be due to a sense of unease or worry that the presence of these children could increase the perception of risks associated with French nuclear installations. Thus, in France, 'Chernobyl children's' holidays have a latent tendency towards politicization, whereas in Britain they contribute to placing radiation health effects in the private sphere.

Finally, the two countries' political systems and cultures demonstrate several interesting similarities and differences that helped to project the respective Chernobyl debates. In both countries, with the exception of The Greens, the public authorities' emergency management of the 1986 crisis did not become a political argument for opposing parties. In France, this can be explained by the fact that the *Première cohabitation* and thereby a broad coalition of the French political spectrum, were responsible for the official reactions to the 1986 crisis. In Britain, however, where one might have expected Labour to have turned the row over sheep farm restrictions into a political argument against the Thatcher government, it seems that Labour's internal disaccord on nuclear policies is what prevented Chernobyl from becoming a political argument at party political level. Other contextual settings and characteristics contributed to the very different development of the French and British Chernobyl debates. The centralized political system in France facilitated the transferral of what began as a local and regional issue to a national topic of debate: the issue of the Chernobyl fallout and its possible health effects did not remain confined to Corsica, Alsace or Mercantour, but quickly reached national proportions. Because most of the key actors in the French Chernobyl debate are present on a national scale – be it the government, the radiation protection agencies (SCPRI, IPSN, IRSN), the plant operator (EDF), or the main anti-nuclear activists (Greenpeace France, Réseau sortir du nucléaire) – their arguments were immediately brought to the national level. In Britain, however, the local and regional debates – in particular regarding sheep farm restrictions in the Lake District and Snowdonia – remained very much on these levels. This is illustrated by the fact that mainly regional, and not national, newspapers reported on this issue, and regional landowners and farmers unions expressed their stance on Chernobyl. Undeniably, applying the term 'regional' to the countries of England, Wales, Scotland and Northern Ireland negates today's constitutional reality. But this is precisely the point I wish to make: the fact that there is a Scottish Parliament with the power to ban and prevent nuclear new build in Scotland has very much diversified the current nuclear debate in the UK. The result is that there is no longer a unitary British nuclear policy today. In France, however, no intermediate power exists that can cross, and question, decisions already made by the national government. All debates on nuclear issues must therefore be taken to the national level. In Britain, the multi-level setting allows for a more scattered or diversified debate. Moreover, the various instruments (inquiries, hearings and committees) that form an important part of the British political decision-making process and procedures, have contributed to creating a British nuclear discourse that is far less confrontational than the one developed in France. This is not to imply that the instrument of inquiries is unproblematic; indeed, it has been severely criticized and accused of being a mere ritual that provides the results the government wants. However, the fact that these inquiries do exist allows for far more public involvement; it at least gives the public a voice in siting decisions. The instrument of committees has also enabled and forced opposing sides within the British nuclear debate to sit round the table and discuss their positions. For instance, the Committee Examining Radiation Risks from Internal Emitters (CERRIE) has brought together Chris Busby and other anti-nuclear activists from Greenpeace and Low Level Radiation Campaign with representatives of the NRPB and BNFL.[9] In France, however, such constellations are far less evident, as the example of Flamanville 3 shows. Opponents of the new build project considered the *enquête publique* (public inquiry) mainly a dishonest attempt to get the public's blessing for a project in which it did not have a real voice.[10] Therefore, many anti-nuclear groups did not take part in the hearings. Despite the broad mobilization against the new build project, a decree by Prime Minister Dominique de Villepin authorized the EDF in April 2007 to go ahead with constructing the first EPR in France.[11] Such events only reflect and reinforce the public perception that the French nuclear techno-political regime is far more closed than its British equivalent.

The French and British Chernobyl debates also demonstrate that a disaster is not debated in a discursive vacuum but is always related and compared to the effects of other disasters. The references to the *affaire du sang contaminé* (contaminated blood scandal) in the French Chernobyl debate and the close connection between Chernobyl and the Sellafield-Windscale complex, which features so prominently in the British Chernobyl debate, are telling examples. At the same time, the respective Chernobyl debates deeply influenced how the actors in both countries reacted to Fukushima. Their reactions exemplify how much the perception and memory of disaster are framed by contemporary settings and previous experiences.

And last but not least, this study not only reveals the high degree of transnationality in the Chernobyl debates, but also how greatly the reception of narratives varied from one national context to the other. The level of success achieved by the transnational proponents of the apocalyptic narrative was dramatically different: in France these actors succeeded in shaping the public debate and how Chernobyl was represented, but in Britain they did not have much success at all. Yuri Bandazhevsky is a good example: while he became a prominent actor in the French Chernobyl debate after 2006, his work does not feature in the British Chernobyl debate. An analysis of the intertextuality between East and West European Chernobyl publications enables us to identify networks of actors, as well as trace the origin of specific arguments and narrative elements. In France, *La supplication* has been particularly popular and thus became an essential part of many French Chernobyl narratives.

My comparison of the French and British Chernobyl debates produced an interesting general picture: different perceptions of the power dynamics within the two national nuclear techno-political regimes explained the different trajectories of the Chernobyl debates. In France, from the outset, Chernobyl was framed as a French debate, and placed in the context of the *nucléocratie*. In Britain, such an interpretative framework did not exist for the civil nuclear programme. The predominant criticism of the nuclear enterprise had always been directed at the military complex and was more focussed on international relations than on national nuclear energy. In Britain, Chernobyl was considered from a global perspective, whereas in France the focus was the accident's impact at home. However, with the end of the Cold War and the British government's decision in the mid-1990s to no longer finance new nuclear power plants, anti-nuclear positions lost their momentum. Thus, few people in Britain were interested in transforming the debate on the health impact of Chernobyl into a fight over the legitimacy of the civil nuclear enterprise as happened in France. Leaving this 'discursive and commemorative gap' to be filled by the solidarity movement, Chernobyl did not become an anti-nuclear *lieu de mémoire* in Britain like in France, and instead became associated with charity activities for disabled or unprivileged children in Eastern Europe, which effectively depoliticized Chernobyl and separated it from the nuclear debate.

Notes

1. This is a quotation from Frewer et al., *Media Reporting*.
2. Hohenemser and Renn, 'Chernobyl's Other Legacy'.
3. English literature often makes a connection between Chernobyl and Revelations 8:10, 11, translating 'Chernobyl' as *wormwood* (as in the plant). This revelation says: 'And the third angel sounded, and there fell a great star from heaven, burning as it were a lamp, and it fell upon a third of the rivers, and upon the springs of water; and the name of the star is called Wormwood: and a third of the waters became bitter; and many people had died of the waters, because they were made bitter'.
4. Debates in Britain might well be held in private or in a non-publicly visible sphere. However, sociology and anthropology methods and entirely different sources would be needed to determine whether such debates exist.
5. Wynne, 'Misunderstood Misunderstandings', 20.
6. *The Telegraph*, 'Only Two in Five Believe Climate Change Caused by Human Activity'. Franz-Josef Brüggemeier's account drew my attention to the British climate change debate, see: Brüggemeier, *Geschichte Großbritanniens im 20. Jahrhundert*.
7. Mackay and Thompson, *Something in the Wind*, 48.
8. Two aspects emerge in the criticism of employing 'nuclear nomads': first, it is considered problematic that French employers may pay less attention to the received radioactive doses of these workers than the doses to which their French unionized colleagues are subjected. Second, there is an underlying fear that these contract workers – often from Eastern Europe – might somehow undermine French security standards.
9. Detailed information on CERRIE was formerly available on the committee's website: http://www.cerrie.org/about.php (last accessed 15 August 2018).
10. For this critique, see: Devisse et al., 'Débat public sur l'EPR'; de Pracontal, 'EPR: premier débat, cahin-caha'.
11. The decree is available at: http://www.legifrance.gouv.fr/affichTexte.do?cidTexte=JORFTEXT000000276348&dateTexte=&categorieLien=id (last accessed 15 February 2020).

Epilogue

If we talk about radiation victims, does it ultimately matter whether someone develops an illness because of the radiation or because of their fear of this radiation? In both cases, the illness is in a way radiation induced. Fear of radiation does exist – and certainly people's fear is often not in line with what professionals in the field of radiation protection consider appropriate. And yes, indeed, fear of radiation is tightly coupled with nuclear technology. But this interrelation did not appear from nowhere; it has a history – and it came to serve very concrete political aims. What is the *force de frappe* other than a means to impose fears? In the context of the Cold War, nuclear fears became a major political instrument and ultimately made the technology – both its (tightly interconnected) military and civil applications – so big and so important. But the fear of nuclear technology is certainly not the only nuclear fear. The pro-nuclear side has prominently appealed to fears for legitimizing the nuclear enterprise: the fear of the political opponent in the Cold War divide, the fear of being left behind in the industrialized world's technology race, the fear of being dependent on Arab oil or the fear of the effects of climate change. The nuclear discourse and its practices are overloaded with all kinds of fears.[1] The academic world could best live up to its reputation by finally starting to question the political economy of these multifaceted fear–technology interrelations, instead of continuously investigating exaggerated risk perception in radiation exposed people.

People will probably never agree on what the health impact of Chernobyl radiation exposure has been, is, or will be. Thus, it is maybe time to focus our attention also on other pressing issues, such as the role academic research has adopted in the Chernobyl debate.

Note

1. Kalmbach, Marklund and Åberg, 'Crises and Technological Futures'.

Bibliography

A2. 'Tchernobyl: rétro des mensonges', *A2 Dernière*, 12 May 1986.
———. 'Interview with Alain Carignon', *MIDI 2*, 11 May 1986.
———. 'Interview with François Cogné', *MIDI 2*, 29 April 1986.
———. 'Interview with Pierre Tanguy', *MIDI 2*, 30 April 1986.
Ackerman, Galia. *Tchernobyl, retour sur un désastre*. Paris: Buchet/Chastel, 2006.
Ackerman, Galia, and Guillaume Grandazzi and Frédérick Lemarchand (eds). *Les silences de Tchernobyl: L'avenir contaminé*. Paris: Editions Autrement, 2006.
Alexievich, Svetlana. *Voices from Chernobyl: Chronicle of the Future*. London: Aurum, 1999.
———. *La supplication: Tchernobyl, chroniques du monde après l'Apocalypse*. Paris: JC Lattès, 1998.
Anders, Günther. *La menace nucléaire: considérations radicales sur l'âge atomique*. Monaco: Le Serpent à Plumes, 2006.
———. *Die atomare Drohung: Radikale Überlegungen zum atomaren Zeitalter*. Munich: C.H. Beck, 1981.
———. *Endzeit und Zeitenende: Gedanken über die atomare Situation*. Munich: C.H. Beck, 1972.
Ardill, John. 'Chernobyl Leaves Shadow over 270,000 Sheep', *The Guardian*, 22 April 1987, p. 2.
Arndt, Melanie (ed.). *Politik und Gesellschaft nach Tschernobyl: (Ost-)Europäische Perspektiven*. Berlin: Christoph Links Verlag, 2016.
——— (ed.). 'Memories, Commemorations, and Representations of Chernobyl', special issue of *Anthropology of East Europe Review*, 30(1) (2012).
———. 'Verunsicherung vor und nach der Katastrophe: Von der Anti-AKW-Bewegung zum Engagement für die "Tschernobyl-Kinder"', *Zeithistorische Forschungen* 7(2) (2010): 240–58.
———. 'Von der Todeszone zum Strahlen-Mekka? Die Erinnerung an die Katastrophe von Tschernobyl in Belarus, der Ukraine und Russland', *Zeitgeschichte-online*, April 2006: https://zeitgeschichte-online.de/sites/default/files/documents/arndt_tschernobyl.pdf (last accessed 15 February 2020).
Baker, Johan. 'Obituary: Lord Marshall of Goring', *The Independent*, 26 February 1996.
Ballard, G.M. (ed.). *Nuclear Safety after Three Mile Island and Chernobyl*. London/New York: Elsevier Applied Science Publishers, 1988.
Bandajevskaya, Galina. 'Comment on a réduit au silence le professor Youri Bandajevski', in Guillaume Grandazzi and Frédérick Lemarchand (eds), *Les silences de Tchernobyl* (Paris: Editions Autrement, 2004), 101–5.

Bandazhevsky, Yuri I. *La philosophie de ma vie: journal de prison*. Paris: Jean-Claude Gawsewitch Editeur, 2006.

Bandazhevsky, Yuri I., and Galina S. Bandajevskaya. *Les conséquences de Tchernobyl sur la santé: Le système cardiovasculaire et l'incorporation de radionucléides Cs-137*. Gap: Editions Yves Michel, 2012.

Bandazhevsky, Yuri I., and N.F. Dubovaya. *Les conséquences de Tchernobyl sur la natalité: Césium radioactif et processus de reproduction*. Gap: Editions Yves Michel, 2012.

Bandazhevsky, Yuri I. et al. *Tchernobyl 25 ans après: Situation démographique et problèmes de santé dans les territoires contaminés*. Gap: Editions Yves Michel, 2011.

Bauer, Susanne, and Karena Kalmbach and Tatiana Kasperki. 'From Pripyat to Paris, from Grassroots Memories to Globalized Knowledge Production: The Politics of Chernobyl Fallout', in Laurel MacDowell (ed.), *Nuclear Portraits* (Toronto: University of Toronto Press, 2017), 149–89.

Beck, Ulrich. 'The Anthropological Shock: Chernobyl and the Contours of the Risk Society', *Berkeley Journal of Sociology: A Critical Review* 32(1987): 153–65.

———. *Risikogesellschaft*. Frankfurt a. M.: Suhrkamp, 1986.

Beesley, Ian. *Orphans of the Fallout: Zhitkovitchi Orphanage Belarus*. London: The Darkroom Press, 2001.

Behrisch, Lars. 'Political Economy and Statistics in the Late Ancien Régime' in Willibald Steinmetz and Ingrid Gilcher-Holtey and Heinz-Gerhard Haupt (eds), *Writing Political History Today* (Frankfurt a. M./New York: Campus, 2013), 175–90.

Belbéoch, Bella. 'Le complot international', *Ecologie* 371 (1986).

Belbéoch, Bella and Roger Belbéoch. *Tchernobyl, une catastrophe: Quelques éléments pour un bilan*. Paris: Ed. Allia, 1993. Available online: http://www.dissident-media.org/info nucleaire/Tchernobyl_une_catastrophe_1993.pdf (last accessed 15 February 2020).

———. 'Tchernobyl, une catastrophe: Quelques éléments pour un bilan', *L'Intranquille, une libre contribution à la critique de la servitude* 1 (1992): 267–373.

Bickerstaff, Karen et al. 'Reframing Nuclear Power in the UK Energy Debate: Nuclear Power, Climate Change and Radioactive Waste', *Public Understanding of Science* 17(2) (2008): 145–69.

Bigot, Bernard. *Point sur les conséquences sanitaires de l'accident de Tchernobyl*. N.p: CEA, 2005. Formerly available at: http://www.cea.fr/energie/accident-de-tchernobyl-reperes/pour-en-savoir-plus (last accessed 15 June 2013).

Bohn, Thomas. 'From Recording the Catastrophe to Tackling the Trauma: Chernobyl in (Post-) Soviet Documentary Literature', in Thomas Bohm et al. (eds), *The Impact of Disaster* (Berlin: EB Verlag, 2015), 147–64.

Bohn, Thomas et al. (eds). *The Impact of Disaster: Social and Cultural Approaches to Fukushima and Chernobyl*. Berlin: EB Verlag, 2015.

Bolouri, Yvonne. 'Nuclear Family of Love', *The Sun*, 25 April 2006.

Bonnett, P.J.P. *Transport Mechanisms and Rates for Long Lived Chernobyl Deposits in Mid-Wales*. N.p.: AEA Technology Environment and Energy, 1991.

Bonß, Wolfgang. *Vom Risiko: Ungewißheit und Unsicherheit in der Moderne*. Hamburg: Hamburger Edition, 1995.

Boudia, Soraya, and Natalie Jas. 'Risk and "Risk Society" in Historical Perspective', special issue of *History and Technology* 23(4) (2007).

Bouvier, Yves. 'Risques perçus et risques industriels: Le nucléaire en France', in Denis Varaschin (ed.), *Risques et prises de risques dans les sociétés industrielles* (Bruselles: Peter Lang, 2007), 139–50.

British Nuclear Energy Society. *Chernobyl: A Technical Appraisal. Proceedings of the Seminar Organized by the British Nuclear Energy Society Held in London on 3 October 1986.* London: Telford for the British Nuclear Energy Society, 1987.

Brown, Jennifer. 'The Impact of Television Coverage on the United Kingdom Public Following the Chernobyl Nuclear Accident', in J.M. Wober (ed.), *Television and Nuclear Power: Making the Public Mind* (Norwood: Ablex Publishing Corporation, 1992).

Brown, Kate. *Manual for Survival: A Chernobyl Guide to the Future*. New York: W.W. Norton, 2019.

Brown, Paul. 'Welsh Sheep Farmers Suffer New Clampdown over Chernobyl Accident', *The Guardian*, 18 September 1987, p. 2.

———. *'Labour at Odds over Nuclear Power'*, *The Guardian*, 1 May 1986, p. 20.

Brüggemeier, Franz-Josef. *Geschichte Großbritanniens im 20. Jahrhundert*. Munich: C.H. Beck, 2010.

Buie, Elizabeth. 'Cancer Island Inquiry Health Board to Probe GPs' Fears of Chernobyl or Sellafield Link to Benbecula Deaths', *The Herald Scotland*, 1 April 1996.

Burke, Patrick. 'END: Transnational Peace Campaigning in the 1980s', in Eckart Conze and Martin Klimke and Jeremy Varon (eds), *Nuclear Threats, Nuclear Fears and the Cold War of the 1980s* (Cambridge: Cambrige University Press, 2017), 227–50.

Bürkner, Daniel. 'Eine vollkommen neue Realität: Transgression des Wahrnehmbaren in den Bildern Tschernobyls', in Ingeborg Reichle and Steffen Siegel, *Maßlose Bilder: Visuelle Ästhetik der Transgression* (Munich: Wilhelm Fink, 2009), 189–206.

Busby, Chris. *The Wings of Death: Nuclear Pollution and Human Health*. Aberystwyth: Green Audit Books, 1995.

Busby, Chris, and Alexey Yablokov (eds). *Chernobyl 20 Years on: Health Effects of the Chernobyl Accident*. Aberystwyth: Published on behalf of the ECRR by Green Audit, 2006.

Busby, Christopher, and Molly Scott Cato. 'Increases in Leukemia in Infants in Wales and Scotland Following Chernobyl: Evidence for Errors in Statutory Risk Estimates', *Energy and Environment* 11(2) (2000): 127–39.

Camplin, W.C., and Great Britain Directorate of Fisheries Research. *Radioactivity in Freshwater Systems in Cumbria (UK) Following the Chernobyl Accident*. N.p.: MAFF, 1989.

Castanier, Corinne. 'Contamination des sols français par les retombées de l'accident de Tchernobyl – les preuves du mensonge', in CRIIRAD and A. Paris (eds), *Contaminations radioactives: atlas France et Europe* (Barret-sur-Méouge: Editions Yves Michel, 2002): 7–27.

Cawse, P.A., and S.J. Baker. *A Survey of Radioactive Caesium in British Soils: Comparison of Accumulations Pre- and Post-Chernobyl*. Harwell: AEA Environment and Energy, 1990.

CEA. 'Accident de Tchernobyl: quelques repères sur le sujet'. Formerly available at: http://www.cea.fr/energie/accident-de-tchernobyl-reperes/reperes (last accessed 15 June 2013).

Chafer, Tony. 'Politics and the Perception of Risk: A Study of the Anti-Nuclear Movements in Britain and France', *West European Politics* 8(1) (1985): 5–23.

Chamberlain, A.C. *Comparisons of the Emissions in the Windscale and Chernobyl Accidents*. Harwell: UKAEA Atomic Energy Research Establishment Environmental and Medical Sciences Division, 1987.

Charpak, Georges, and Richard L. Garwin and Venance Journé. *De Tchernobyl en tchernobyls.* Paris: Odile Jacob, 2005.

Chernousenko, Vladimir Mikhailovich. *Chernobyl: Insight from the Inside.* Berlin/London: Springer, 1991.

Chiquelin, Jacques. 'Le clan de l'atome français', *Libération*, 13 May 1986.

Clark, Michael. 'John Dunster CB ARCS BSc FSRP (1922–2006)'. Formerly available at: http://www.hpa.org.uk/Publications/Radiation/MiscellaneousRadiationPublications/rad20misc_pub_DunsterObituary/ (last accessed 15 June 2013).

Clarke, Robert. 'Le Matin répond aux 10 questions que vous vous posez', *Le Matin*, 2 May 1986.

Clouston, Erlend. 'Legacy of Night when it Rained Radiation', *The Guardian*, 26 April 1996, p. 6.

CND. 'Sellafield, Libya, Chernobyl', *The Guardian*, 7 May 1986, p. 10.

———. 'What if Nuclear War was just an Accident', *The Guardian*, 2 May 1986, p. 27.

———. 'The History of CND': http://www.cnduk.org/about/item/437 (last accessed 15 February 2020).

Collier, John G., and Myrddin L. Davies. *Chernobyl.* Gloucester: Central Electricity Generating Board, 1986.

Collins, Harry, and Trevor Pinch. *Der Golem der Technologie: Wie unsere Wissenschaft die Wirklichkeit konstruiert.* Berlin: Berlin Verlag, 2000.

———. *The Golem at Large: What you Should Know about Technology.* Cambridge: Cambridge University Press, 1998.

Condon, Judith. *Chernobyl and Other Nuclear Accidents.* Hove: Wayland, 1998.

Cook, Stephen. 'Render unto Caesium 134', *The Guardian*, 1 June 1990, p. 23.

Cotterill, S.J., and M.S. Pearce and L. Parker. 'Thyroid Cancer in Children and Young Adults in the North of England: Is Increasing Incidence Related to the Chernobyl Accident?', *European Journal of Cancer* 37(8) (2001): 1020–26.

Country Landowners' Association North Wales Group. *North Wales Group Report on the Effects of the Chernobyl Accident.* N.p.: 1987.

Crié, Hélène. 'La France, lanterne rouge européenne de l'information', *Libération*, 9 May 1986.

Crié, Hélène, and Yves Lenoir. *Tchernobyl-sur-Seine.* Paris: Calmann-Lévy, 1987.

CRIIRAD. *Tchernobyl – 20 ans après, les services officiels français persistent dans la censure et la désinformation.* Valence: CRIIRAD, 2006.

———. *Trait d'Union: Bulletin d'information des adhérents de la CRIIRAD.* No 32/33. N.p.: 2005.

CRIIRAD, and AFMT. *Communiqué joint: Mise en examen du Pr Pellerin pour tromperie aggravée dans le dossier Tchernobyl / malades de la thyroïde.* N.p.: 1 June 2006.

CRIIRAD, and André Paris (eds). *Contaminations radioactives: atlas France et Europe.* Barret-sur-Méouge: Editions Yves Michel, 2002.

Croizé, Jean-Paul. 'Le syndrome ukrainien', *Le Figaro*, 30 April 1986.

Cunningham, John. 'When Lamb Provides the Beef', *The Guardian*, 23 June 1986, p. 21.

Cutler, James, and Rob Edwards. *Britain's Nuclear Nightmare.* London: Sphere, 1988.

David-Jougneau, Maryvonne. 'Semmelweis, Bandajevsky: des savants victimes de la répression scientifique', in Guillaume Grandazzi and Frédérick Lemarchand (eds), *Les silences de Tchernobyl* (Paris: Editions Autrement, 2004), 106–18.

Department of Agriculture and Fisheries for Scotland. *Chernobyl Accident: Monitoring for Radioactivity in Scotland*. N.p.: Department of Agriculture and Fisheries for Scotland, 1990.

Department of the Environment (Her Majesty's Inspectorate of Pollution). *Nuclear Accidents Overseas: The National Response Plan and Radioactive Incident Monitoring Network (RIMNET). A Statement of Proposals*. London: HMSO, 1988.

Department of the Environment et al. *Levels of Radioactivity in the UK from the Accident at Chernobyl USSR, on 26 April 1986*. London: HMSO, July 1986.

de Pracontal, Michel. 'EPR: premier débat, cahin-caha', *Le Nouvel Observateur*, 10 November 2005.

Desbordes, Jean-Philippe. *Atomic Park: À la recherche des victimes du nucléaire*. Arles: Actes Sud, 2006.

Devisse, Jean-Stéphane et al. 'Débat public sur l'EPR: une crédibilité à construire', *Les Echos*, 10 January 2005.

Downer, John. *Nuclear Safety: A (Charlie) Brownian Notion*, Working Paper No. 07-12, University of Bristol, School of Sociology, Politics and International Studies: http://www.bristol.ac.uk/global-insecurities/workingpapers/downer.pdf (last accessed 15 February 2020).

Dowswell, Paul. *The Chernobyl Disaster: 26th April 1986*. London: Hodder Wayland, 2003.

Dupuy, Jean-Pierre. *Retour de Tchernobyl: journal d'un homme en colère*. Paris: Editions du Seuil, 2006.

———. *Pour un catastrophisme éclairé: quand l'impossible est certain*. Paris: Editions du Seuil, 2002.

East, B.W. *Measurement of Radioactivity from Chernobyl in Population Groups in Scotland: Report 2*. N.p.: Her Majesty's Inspectorate of Pollution, 1991.

EDF. *Rapport annuel 2008*. Paris: Groupe EDF, 2009.

———. *Tchernobyl: le vrai, le faux et l'incertain*. Paris: EDF, 1992.

———. *Tchernobyl*. Dijon: EDF, 1987.

———. *Tchernobyl. Dossier d'information*. N.p: EDF, 1986.

Edmondson, B. 'The Chernobyl Accident and Its Implications for Operators of Civil Nuclear Power Plants in the UK', in *Nuclear Risks: Reassessing the Principles and Practice after Chernobyl. Conference: Papers and Discussions*. London: IBC, 1987.

Edwards, Rob. 'Chernobyl Fall-Out "40 Times Worse" than Admitted', *The Guardian*, 28 January 1989, p. 24.

Edwards, Rob, and Susan D'Arcy. *Still Fighting for Gemma*. London: Bloomsbury, 1995.

Enfants de Tchernobyl Bélarus. 'La plaquette réactualisée de l'association'. Formerly available at: http://enfants-tchernobyl-belarus.org/doku.php (last accessed 15 November 2013).

———. 'Compte rendu de l'Assemblée Générale Ordinaire du 17 novembre 2012'. Saumur: ETB, 2012. Available online: http://enfants-tchernobyl-belarus.org/lib/exe/fetch.php?media=archives:rapports_internes:cr_ag_11-2011.pdf (last accessed 15 February 2020).

European Commission. *Atlas on the Caesium Deposition across Europe after the Chernobyl Accident*. Luxembourg: Office for Official Publications of the European Communities, 1996.

Fairhall, David. 'Chernobyl Factors', *The Guardian*, 30 April 1986, p. 23.

———. 'Radioactive Russian Dust Cloud Escapes', *The Guardian*, 29 April 1986, p. 1.

Fairhall, David, and Michael White and Mark Tran. 'Radiation Cloud Reaches Britain', *The Guardian*, 3 May 1986, p. 1.
Fairlie, Ian, and David Sumner. *The Other Report on Chernobyl (TORCH)*. Berlin/Brussels/Kiev: The Greens and EFA in the European Parliament, 2006.
Farmers' Union of Wales. *Chernobyl: The Government Reaction*. N.p.: Farmers' Union of Wales, 1988.
FEFU. 'Portrait': http://www.fefu.org/default.asp?voirpage=FEFU/Portrait.htm (last accessed 15 February 2020).
Fischhoff, Baruch. 'The Nuclear Energy Industry's Communication Problem', *Bulletin of the Atomic Scientist* (2009): http://thebulletin.org/nuclear-energy-industrys-communication-problem (last accessed 15 February 2020).
———. 'Risk Perception and Communication Unplugged: Twenty Years of Process', *Risk Analysis* (15)2 (1995): 137–45.
Forgrave, Andrew. 'Wales: Did Chernobyl Disaster Cause my Cancer?', *Daily Post*, 26 April 2006.
Fox, Robert. 'Obituary: Professor Margaret Gowin', *The Independent*, 20 November 1998: http://www.independent.co.uk/arts-entertainment/obituary-professor-margaret-gowing-1186010.html (last accessed 15 February 2020).
France 3. *Soir 3*, 28 April 1986.
Freudenberg, William R. 'Perceived Risk, Real Risk: Social Science and the Art of Probabilistic Risk Assessment', *Science* 242 (1988): 44–9.
Frevert, Ute. *Vertrauensfragen: Eine Obsession der Moderne*. Munich: C.H. Beck, 2013.
Frewer, Lynn et al. *Media Reporting of Risk in 5 European Countries at the Time of the 10th Anniversary of the Chernobyl Accident, Report to the European Commission*. Norwich: Institute of Food Research, 1997.
Friends of the Earth. *The Chernobyl Legacy*. London: Friends of the Earth, 1991.
———. 'A Positive Message on Chernobyl from Friends of the Earth', *The Observer*, 11 May 1986, p. 15.
———. 'The Accident They Said Would Never Happen', *The Guardian*, 6 May 1986, p. 31.
Gale, Robert Peter. *Chernobyl: The Final Warning*. London: Hamilton, 1988.
German Affiliate of International Physicians for the Prevention of Nuclear War (IPPNW), and Gesellschaft für Strahlenschutz. *Health Effects of Chernobyl: 25 Years after the Reactor Catastrophe*. Berlin: IPPNW / Gesellschaft für Strahlenschutz, April 2011.
Ghazi, Polly. 'Chernobyl Fall-Out May Affect British Farmland for Decades', *The Observer*, 29 April 1990, p. 9.
Gittus, John et al. *The Chernobyl Accident and Its Consequences*. London: United Kingdom Atomic Energy Authority, 1987.
Goatcher, Jeff, and Viv Brunsden. 'Chernobyl and the Sublime Tourist', *Tourist Studies* 11(2) (2011): 115–37.
Gould, Peter. *Fire in the Rain: The Democratic Consequences of Chernobyl*. Cambridge: Polity Press, 1990.
Gowing, Margaret, assisted by Lorna Arnold. *Independence and Deterrence: Britain and Atomic Energy, 1945–1952*. London: Macmillan, 1974.
Gowing, Margaret (in collaboration with Lorna Arnold). *Britain and Atomic Energy, 1935–1945*. London: Macmillan, 1964.

Grandazzi, Guillaume. 'Die Zukunft erinnern – Gedenken an Tschernobyl', *Osteuropa*, 56(4) (2006): 7–18.

———. 'L'atome en héritage', in Guillaume Grandazzi and Frédérick Lemarchand (eds), *Les silences de Tchernobyl* (Paris: Editions Autrement, 2004), 120–30.

Grandazzi, Guillaume, and Frédérick Lemarchand (eds). *Les silences de Tchernobyl: L'avenir contaminé*. Paris: Editions Autrement, 2004.

———. 'Avant-propos', in Guillaume Grandazzi and Frédérick Lemarchand (eds), *Les silences de Tchernobyl* (Paris: Editions Autrement, 2004), 5–11.

Greenpeace. *The Chernobyl Catastrophe: Consequences on Human Health*. Amsterdam: Greenpeace, 2006.

Greenpeace France. *20 ans Tchernobyl*. Paris: Greenpeace France, 2006.

Greenpeace UK. 'Hinkley Strike Price Briefing'. Formerly available at: http://www.greenpeace.org.uk/hinkley-strike-price-briefing (last accessed 15 June 2013).

Guihannec, Yves. 'Nucléaire: les silences de la France', *l'Express*, 16 May 1986.

Halsall, Martyn. 'New Controls Spur Fall-Out Inquiry Call', *The Guardian*, 15 August 1987, p. 3.

Hamman, Henry, and Stuart Parrott. *Mayday at Chernobyl*. Sevenoaks: New English Library, 1987.

Haynes, Viktor, and Marko Bojcun. *The Chernobyl Disaster*. London: Hogarth, 1988.

Haywood, J.K., and Institute of Physical Sciences in Medicine. *Chernobyl: Response of Medical Physics Departments in the United Kingdom*. London: Institute of Physical Sciences in Medicine, 1986.

Health Protection Agency. 'The Consequences of the Chernobyl Nuclear Accident', in: HPA Press Releases Archive for 2006. Formerly available at: http://www.hpa.org.uk/webc/HPAwebFile/HPAweb_C/1253205582795 (last accessed 15 June 2013).

Heath, Tony. 'Nuclear Cloud Hangs over the Hills', *The Independent*, 22 April 1996.

———. 'Radiation Checks on People in North Wales', *The Guardian*, 28 September 1987, p. 4.

———. 'Farmers Take Chernobyl "Hostages"', *The Guardian*, 5 September 1986, p. 3.

———. 'Contaminated Rain Sets Welsh Alarms Ringing', *The Guardian*, 7 May 1986, p. 2.

Hecht, Gabrielle. *Being Nuclear: Africans and the Global Uranium Trade*. Cambridge, MA: MIT Press, 2012.

———. 'Nuclear Ontologies', *Constellations* 13(3) (2006): 320–31.

———. *The Radiance of France: Nuclear Power and National Identity after World War II*. Cambridge, MA: MIT Press, 1998.

Heuser, Beatrice. *Nuclear Mentalities? Strategies and Belief-Systems in Britain, France and the FRG*. Basingstoke: Palgrave Macmillan, 1998.

Hickman, Leo. 'What Is Safe to Eat?', *The Guardian*, 29 December 2009.

Highfield, Roger. 'Windscale Fallout from 1957 Reactor Fire still Affecting Cumbria', *The Daily Telegraph*, 25 April 2006.

Hindmarsh, Richard (ed.). *Nuclear Disaster at Fukushima Daiichi: Social, Political and Environmental Issues*. New York: Routledge, 2013.

Hogg, Jonathan, and Christoph Laucht (eds). 'British Nuclear Culture', special issue of *The British Journal for the History of Science* 45(4) (2012).

Hohenemser, Christoph, and Ortwin Renn. 'Chernobyl's Other Legacy: Shifting Public Perceptions of Nuclear Risk', *Environment* 30(3) (1988): 5–11 and 40– 45.

Hooper, John. 'Britain Sticks to Policy', *The Guardian*, 1 May 1986, p. 8.
House of Commons Agricultural Committee. *Chernobyl – The Government's Response, Supplementary Memorandum by the MAFF*. N.p.: June 1988.
Hull Daily Mail. 'We Won't Forget Chernobyl Victims', 25 April 2006.
Hunt, Liz. 'Chernobyl Link to Cancer Cluster on Scottish Island', *The Independent*, 31 March 1996.
IBC Technical Service. *Nuclear Risks: Reassessing the Principles and Practice after Chernobyl. Conference: Papers and Discussions*. London: IBC, 1987.
Independent on Sunday. 'Chernobyl "Still Causing Cancer in British Children"', 23 April 2006.
IndependentWHO. 'The Collective IndependentWHO': http://independentwho.org/en/the-collective-independentwho/ (last accessed 15 February 2020).
Ingram, Scott. *The Chernobyl Nuclear Disaster*. New York/London: Facts on File, 2005.
Internationales Bildungs- und Begegnungswerk Dortmund (ed.). *Tschernobyl und die europäische Solidaritätsbewegung*. Norderstedt: Books on Demand, 2011.
InVS (L. Pascal, J.L. Lasalle). *Estimation de l'incidence du cancer de la thyroïde en Corse. 1998 – 2006*. Saint-Maurice: InVS, 2012.
InVS (Laurence Chérié-Challine et al.). *Surveillance sanitaire en France en lien avec l'accident de Tchernobyl – Bilan actualisé sur les cancers thyroïdiens et études épidémiologiques en cours en 2006*. Saint-Maurice: InVS, 2006.
IPSN, and InVS (P. Verger, L. Chérié-Challine). *Evaluation des conséquences sanitaires de l'accident de Tchernobyl en France*. Saint-Maurice: InVS, 2001.
———. *Evaluation des conséquences sanitaires de l'accident de Tchernobyl en France – synthèse du rapport*. Saint-Maurice: InVS, 2001.
IPSN (Ph. Renaud, K. Beaugelin, H. Maubert, Ph. Ledenvic). *Conséquences radioécologiques et dosimétriques de l'accident de Tchernobyl en France*. Fontenay-aux-Roses: IPSN, 1997.
IPSN. *Tchernobyl, 10 ans après — dossier de presse*. Fontenay-aux-Roses: IPSN-Mission communication, 1996.
IRSN. *Le Baromètre IRSN sur la perception des risques et de la sécurité par les Français – Les essentiels*. Fontenay-aux-Roses: IRSN, 2018.
———. *Baromètre IRSN 2006: La perception des situations à risques par les Français*. Fontenay-aux-Roses: IRSN, 2006.
———. 'Les leçons de Tchernobyl'. Formerly available at: http://www.irsn.org/index.php position=lecons_tchernobyl_accueil (last accessed June 2009). A new dossier was published for the thirtieth anniversary: http://www.irsn.fr/FR/connaissances/Installations_nucleaires/Les-accidents-nucleaires/accident-tchernobyl-1986/Pages/Tchernobyl.aspx (last accessed 15 February 2020).
———. 'Les leçons de Tchernobyl – Evaluer les conséquences et protéger les personnes'. Formerly available at: http://www.irsn.org/index.php?position=lecons_tchernobyl_la_mortalite_due_a_l_accident (last accessed June 2009).
———. 'Les leçons de Tchernobyl – Exposition des personnes en France'. Formerly available at: http://www.irsn.org/index.php?position=lecons_tchernobyl_exposition_des_person nes_en_france (last accessed June 2009).
———. 'Les leçons de Tchernobyl – Mieux agir en territoire contaminé'. Formerly available at: http://www.irsn.org/index.php?position=mieux_agir_en_territoire_contamine (last accessed June 2009).

Jacquemin-Raffestin, Jean-Michel. *Tchernobyl 20 ans après: Cachez ce nuage que je ne saurais voir*. Paris: Guy Trédaniel Editeur, 2006.

———. *Tchernobyl, conséquences en France. J'accuse…!*. Paris: Editions Sang de la Terre, 2002.

———. *Tchernobyl: Aujourd'hui les Français malades*. Monaco: Editions du Rocher, 2001.

Jacquemin, Jean-Michel. *Ce fameux nuage … Tchernobyl, la France contaminée*. Paris: Editions Sang de la Terre, 1999.

———. *Ce fameux nuage … Tchernobyl, la France contaminée*. Paris: Editions Sang de la Terre, 1998.

Jamet, Dominique. 'Le S.O.S de l'URSS', *Le Quotidien de Paris*, 30 April 1986.

Johnson, Angella. 'Walker Pledges Review of Chernobyl Claims', *The Guardian*, 29 September 1987, p. 3.

Johnson, Thomas. *La bataille de Tchernobyl*. Paris: Play Film, 2006.

Johnstone, Helen. 'Health Director to Study Big Rise in Benbecula Cancer', *The Times*, 1 April 1996.

Jopling, Michael. 'Sheep Still too Hot to Handle', *The Guardian*, 7 March 1987, p. 11.

Jordan, Katrin. *Ausgestrahlt: Die mediale Debatte um 'Tschernobyl' in der Bundesrepublik und in Frankreich 1986/87*. Göttingen: Wallstein, 2018.

Jowit, Juliette. 'Dispatch: Chernobyl's Generations of Suffering', *The Observer*, 23 April 2006.

Kalmbach, Karena. 'Point de vue', in IRSN (ed.), *Le Baromètre IRSN sur la perception des risques et de la sécurité par les Français – Les essentiels* (Fontenay-aux-Roses: IRSN, 2018), p. 24. Available online: http://barometre.irsn.fr/essentiels2018/#p=26 (last accessed 15 February 2020).

———. 'Tchernobyl – angle mort', in Etienne François and Thomas Serrier (eds), *Europa notre histoire: L'héritage européen depuis Homère* (Paris: Les Arènes, 2017), 316–18.

———. 'Revisiting the Nuclear Age: State of the Art Research in Nuclear History', *Neue Politische Literatur* 62(1) (2017): 49–69.

———. 'Frankreich nach Tschernobyl: Eine Rezeptionsgeschichte zwischen "Nicht-Ereignis" und "Apokalypse"', in Melanie Arndt (ed.), *Politik und Gesellschaft nach Tschernobyl: (Ost-)Europäische Perspektiven* (Berlin: Christoph Links Verlag, 2016), 237–55.

———. 'Meanings of a Disaster: The Contested "Truth" about Chernobyl. British and French Chernobyl Debates and the Transnationality of Arguments and Actors', PhD dissertation. Florence: European University Institute, 2014. Available online: https://cadmus.eui.eu/handle/1814/33491 (last accessed 15 February 2020).

———. 'Von Strahlen und Grenzen: Tschernobyl als nationaler und transnationaler Erinnerungsort', in Frank Uekötter (ed.), *Ökologische Erinnerungsorte* (Göttingen: Vandenhoeck & Ruprecht, 2014), 186–215.

———. 'Radiation and Borders: Chernobyl as a National and Transnational Site of Memory', *Global Environment* 11(2013): 130–59.

———. 'Die Wolke, die an der Grenze haltmachte: Zur Wahrnehmung der Auswirkungen von Tschernobyl in Frankreich', in IBB (ed.), *Tschernobyl und die europäische Solidaritätsbewegung* (Norderstedt: Books on Demand, 2011), 74–88.

———. *Tschernobyl und Frankreich: Die Debatte um die Auswirkungen des Reaktorunfalls im Kontext der französischen Atompolitik und Elitenkultur*. Frankfurt a. M.: Peter Lang Verlag, 2011).

Kalmbach, Karena, Andreas Marklund and Anna Åberg. 'Crises and Technological Futures: Experiences, Emotion and Action', *Technology and Culture* 61(1) (2020): 272–281.

Kane, Paul. *Chernobyl: Evaluation and Radiological Assessment of Factors Affecting Drinking Water Supplies*. N.p.: Her Majesty's Inspectorate of Pollution, 1989.

Kasperski, Tatiana. 'La politique de la mémoire d'une catastrophe nucléaire: les usages de l'accident de Tchernobyl en Biélorussie (1986–2008)', PhD dissertation. Paris: Sciences Po Paris, 2012.

Keel, Paul, and Martin Walker. 'British Evacuees Fly Home to Health Checks', *The Guardian*, 1 May 1986, p. 1.

Kirchhof, Astrid Mignon, and Jan-Henrik Meyer. 'Global Protest against Nuclear Power: Transfer and Transnational Exchange in the 1970s and 1980s', *Historical Social Research* 39(1) (2014): 165–90.

Kostin, Igor. *Chernobyl: Confessions of a Reporter*. New York: Umbrage Editions, 2006.

Kostin, Igor (in collaboration with Thomas Johnson). *Tchernobyl: confessions d'un reporter*. Paris: Editions des Arènes, 2006.

Kuchinskaya, Olga. 'Twice Invisible: Formal Representations of Radiation Danger', *Social Studies of Science* 43(1) (2012): 78–96.

Lashmer, Paul, and Robin McKie and Geoffrey Lean. 'Confusion in Britain', *The Guardian*, 11 May 1986, p. 11.

Laucht, Christoph. *Elemental Germans: Klaus Fuchs, Rudolf Peierls and the Making of British Nuclear Culture 1939-59*. Basingstoke: Palgrave Macmillan, 2012.

Lauvergeon, Anne, and Michel-Hubert Jamard. *La troisième révolution énergétique*. Paris: Plon, 2008.

Le Figaro. 'Tchernobyl: mort du Professeur Pierre Pellerin', 3 March 2013: http://www.lefigaro.fr/actualite-france/2013/03/03/01016-20130303ARTFIG00110-tchernobyl-mort-du-professeur-pierre-pellerin.php? (last accessed 15 February 2020).

Le Monde. 'La catastrophe de Tchernobyl pourrait être à l'origine de 24 000 décès par cancer', 28 August 1986, p. 20.

Le Parisien. 'Le nucléaire en France: la sécurité avant tout', 30 April 1986.

Le Quotidien de Paris. 'Catastrophe nucléaire en URSS', 29 April 1986.

Lean, Geoffrey. 'Nuclear Family Hunted by a Testing Legacy', *The Observer*, 29 June 1986, p. 5.

Lean, Geoffrey, and Tony Heath and David Siddell. 'Fell Lambs Born under N Cloud', *The Observer*, 3 May 1987, p. 5.

Lehtonen, Markku. 'Reactions to Fukushima in Finland, France and the UK – Rupture or Continuity in the Nuclear Techno-Politics?', *Second 3.11 Virtual Conference (2013)*: http://fukushimaforum.wordpress.com/online-forum-2/second-3-11-virtual-conference-2013/reactions-to-fukushima-in-finland-france-and-the-uk-rupture-or-continuity-in-the-nuclear-techno-politics/ (last accessed 15 February 2020).

Lehtonen, Markku, and Claire Le Renard and Arthur Jobert. 'The Diverging Trajectories of Fast Breeder Reactor Development in France and the UK (1950s–1990s): A Tentative Comparison', paper presented at the 6th Tensions of Europe plenary conference *Democracy and Technology: Europe in Tension from the 19th to the 21st century*, Paris, 19–21 September 2013.

Lemarchand, Frédérick et al. (eds). *Destination Tchernobyl*. Paris: Editions Syllepse, 2006 (Ecologie & Politique 32/2006).

Les Annales des Mines. 'Pierre Lucien Jean Guillaumat (1909-1991)': http://www.annales.org/archives/x/guillaumat.html (last accessed 15 February 2018).

Les Dernières Nouvelles d'Alsace. 'Polémique sur l'information nucléaire en France', 12 May, 1986.

Lhomme, Stéphane. *L'insécurité nucléaire: bientôt un Tchernobyl en France?*. Barret-sur-Méouge: Editions Yves Michel, 2006.

Libération. 'Message aux malades de la thyroïde ...', 19 November 2005, p. 10.

———. 'L'Europe dans le nuage', 3 and 4 May 1986.

Liberatore, Angela. *The Mangement of Uncertainty: Learning from Chernobyl*. Amsterdam: Gordon and Breach Publishers, 1999.

Low Level Radiation Campaign. *Do We Really Want Nine Chernobyl Accidents Every Year for the Next Ten Years? European Council Directive 96/29/Euratom the So-Called "Basic Standards Directive" on Radiation Protection*. N.p.: Low Level Radiation Campaign, 1999.

Luhmann, Niklas. *Risk: A Sociological Theory*. Berlin/New York: de Gruyter, 1993.

MacDonald, Calcun. 'Isle of Desolate Beauty Haunted by the Spectre of Chernobyl', *The Herald Scotland*, 1 April 1996.

MacDonald, Eileen. 'High Radiation Levels Scar Rural Families', *The Observer*, 23 August 1987, p. 6.

Mackay, Louis, and Mark Thompson (eds). *Something in the Wind: Politics after Chernobyl*. London: Pluto, 1988.

Marples, David. *Belarus: From Soviet Rule to Nuclear Catastrophe*. Basingstoke: Macmillan, 1996.

———. *The Social Impact of the Chernobyl Disaster*. Basingstoke: Macmillan, 1988.

———. *Chernobyl and Nuclear Power in the USSR*. Basingstoke: Macmillan in association with Canadian Institute of Ukrainian Studies, University of Alberta, 1987.

Marshall, Walter (The Lord Marshall of Goring). *The Chernobyl Accident*. London: CEGB, December 1986.

McKie, Robin. 'Britain's Deadly Chernobyl Legacy', *The Observer*, 21 April 1991, p. 11.

McRae, Hamish. 'The Soviet Nuclear Disaster Seems Likely to Put King Coal back on the Throne', *The Guardian*, 30 April 1986, p. 24.

Medvedev, Grigori. *No Breathing Room: The Aftermath of Chernobyl*. New York: Basic Books, 1993.

———. *The Truth about Chernobyl*. London: I.B. Tauris, 1991.

———. *La Vérité sur Tchernobyl*. Paris: A. Michel, 1990.

Medvedev, Zhores A. *The Legacy of Chernobyl*. Oxford: Basil Blackwell, 1990.

———. *Désastre nucléaire en Oural*. Cherbourg: Isoète, 1988.

———. *Nuclear Disaster in the Urals*. New York: Norton, 1979.

Meikle, James. 'Sheep Farms under Curbs See No End to Chernobyl Fallout', *The Guardian*, 13 April 2006.

———. 'Chernobyl Legacy Lingers down on the Farm', *The Guardian*, 1 May 1999, p. 10.

Micoulaut, Raymond. *Tchernobyl: L'histoire d'une désinformation*. Paris: L'Harmattan, 2006.

Ministère de l'Economie, des Finances et de l'Industrie. *Une pièce essentielle dans l'organisation de la sûreté nucléaire et de la radioprotection: le nouvel IRSN. Extrait du rapport annuel 2001 de la DGEMP*. Paris: 2002.

Ministry of Agriculture, Fisheries and Food. *Radionuclide Levels in Food, Animals and Agricultural Products 1987: Post Chernobyl Monitoring in England and Wales.* London: HMSO, 1988.

Mizzi, Victor. 'Über die Arbeit von Chernobyl Children's Lifeline', in IBB (ed.), *Tschernobyl und die europäische Solidaritätsbewegung* (Norderstedt: Books on Demand, 2011), 105–7.

Monbiot, George. 'Christopher Busby's Wild Claims Hurt Green Movement and Green Party', *The Guardian Online*, 22 November 2011: http://www.guardian.co.uk/environment/georgemonbiot/2011/nov/22/christopher-busby-nuclear-green-party (last accessed 15 February 2020).

———. 'Evidence Meltdown', 4 April 2011: http://www.monbiot.com/2011/04/04/evidence-meltdown/ (last accessed 15 February 2020).

Monbiot, George et al. 'A Letter to David Cameron', 15 March 2012: http://www.monbiot.com/2012/03/15/a-letter-to-david-cameron/ (last accessed 15 February 2020).

Montellier, Chantal. *Tchernobyl mon amour*. Arles: Actes Sud BD, 2006.

Morel, Michel. 'L'état français a déjà fait le choix du tout-nucléaire', *Rouge*, 15 May 1986.

Mould, Richard F. *Chernobyl Record: The Definitive History of the Chernobyl Catastrophe.* Bristol: Institute of Physics Publishing, 2000.

———. *Chernobyl: The Real Story*. Oxford: Pergamon Press, 1988.

Mühlenhöver, Emmanuelle. *L'environnement en politique étrangère. Raisons et illusions: une analyse de l'argument environnemental dans les diplomaties électronucléaires française et américaine*. Paris: L'Harmattan, 2002.

Murtagh, Peter. 'Soviet Reactor "Safer than British AGR"', *The Guardian*, 8 May 1986, p. 2.

Naughtie, James, and David Rose. 'MPs Told: There Is No Health Risk', *The Guardian,* 7 May 1986, p. 1.

Nehring, Holger. 'Cold War, Apocalypse, and Peaceful Atoms: Interpretations of Nuclear Energy in British and West German Anti-Nuclear Weapons Movements', *Historical Social Research* 29(3) (2004): 150–70.

Nelkin, Dorothy, and Michael Pollak. *The Atom Besieged: Antinuclear Movements in France and Germany*. Cambridge, MA: MIT Press, 1982.

Nesterenko, Vassilli. 'L'Europe aurait pu devenir inhabitable', in Guillaume Grandazzi and Frédérick Lemarchand (eds), *Les silences de Tchernobyl* (Paris: Editions Autrement, 2004), 14–27.

Nisbet, Anne. 'Management of Chernobyl-Restricted Areas in England and Wales', *Radiological Protection Bulletin* 211(April 1999): 11–15.

Nisbet, Anne S., and R.F.M. Woodman. *Options for the Management of Chernobyl-Restricted Areas in England and Wales.* N.p.: NRPB, 1998.

NRPB. 'Response to Wind Blown Particles and Cancer Mortality'. Formerly available at: http://www.hpa.org.uk/webc/HPAwebFile/HPAweb_C/1194947375682 (last accessed 15 November 2013).

———. *A Preliminary Assessment of the Radiological Impact of the Chernobyl Reactor Accident on the Population of the European Community*. Luxembourg: Commission of the European Communities, 1987.

O'Connell, Vince. 'Letter to the Editor', *The Guardian*, 3 May 1986, p. 12.

O'Meara, Liam. *Fallout: The Children of Belarus and the People of Ireland after Chernobyl.* Dublin: Columba Press, 2003.

Onaga, Lisa. 'Reconstructing the Linear No-Threshold Model in Japan: A Historical Perspective on the Technics of Evaluating Radiation Exposure', *Technology and Culture* 58(1) (2017): 194–205.

O'Riordan, Mike. *Radiation Protection: A Memoir of the National Radiological Protection Board*. Didcot: Health Protection Agency, 2007.

Page, Campbell. '"Information Withheld" on Radioactivity in France', *The Guardian*, 12 May 1986.

Page, Jeremy. 'Chernobyl Hero Remembers the Men Who Saved Europe – Factbox', *The Times*, 22 April 2006.

Park, Chris C. *Chernobyl: The Long Shadow*. London: Routledge, 1989.

Parker, Victoria. *Chernobyl 1986: An Explosion at a Nuclear Power Station*. Oxford: Raintree, 2006.

Perrow, Charles. *Normal Accidents: Living with High-Risk Technologies*. Princeton: Princeton University Press, 1984.

Petryna, Adriana. *Life Exposed: Biological Citizens after Chernobyl*. Princeton: Princeton University Press, 2002.

Phillips, Sarah Drue. 'Chernobyl's Sixth Sense: The Symbolism of an Ever-Present Awareness', *Anthropology and Humanism*, 29(2) (2004): 159–85.

Read, Piers Paul. *Ablaze: The Story of Chernobyl*. London: Mandarin, 1995.

Renaud, Philippe et al. *Les retombées en France de l'accident de Tchernobyl: conséquences radioécologiques et dosimétriques*. Les Ulis: EDP sciences, 1999.

Renaud, Philippe, and Didier Champion and Jean Brenot. *Les retombées radioactives de l'accident de Tchernobyl sur le territoire français: Conséquences environnementales et expositions des personnes*. Paris: Editions Tec & Doc, 2007.

Renn, Ortwin. 'Public Responses to the Chernobyl Accident', *Journal of Environmental Psychology*, 10(2) (1990): 151–67.

Richard, Pascale. 'Tchernobyl: Quelles retombées en France?', *Le Quotidien de Paris*, 23 and 24 August 1986.

Rickard, Graham. *The Chernobyl Catastrophe*. Hove: Wayland, 1988.

Rivasi, Michèle, and Hélène Crié. *Ce nucléaire qu'on nous cache*. Paris: Albin Michel, 1998.

Roche, Adi. *Children of Chernobyl: The Human Cost of the World's Worst Nuclear Disaster*. London: Fount, 1996.

Rootes, Christopher. 'Britain', in Christopher Rootes (ed.), *Environmental Protest in Western Europe* (Oxford: Oxford University Press, 2003), 20–58.

Roussel, Violaine. 'Scandales et redéfinitions de la responsabilité politique. La dynamique des affaires de santé et de sécurité publiques', *Revue française de science politique* 58(6) (2008): 953–83.

Rüdig, Wolfgang. 'Mobilising Environmental Expertise: "Alternative" Research Institutes in Germany'. University of Strathclyde Working Paper, 2000.

Ryder, E.A. 'Regulation of Nuclear Power in the UK after Chernobyl', in *Nuclear Risks: Reassessing the Principles and Practice after Chernobyl: Conference: Papers and Discussions*. London: IBC, 1987.

Sahm, Astrid. 'Auf dem Weg in eine transnationale Gesellschaft? Belarus und die internationale Tschernobyl-Hilfe', *Osteuropa*, 56(4) (2006): 105–16.

———. *Transformation im Schatten von Tschernobyl: Umwelt- und Energiepolitik im gesellschaftlichen Wandel von Belarus und Ukraine*. Münster: LIT Verlag, 1999.

———. 'Und der dritte Weltkrieg heißt Tschernobyl ...', in F. Dorn et al. (eds), *Erinnerungen gegen den Krieg* (Minsk: Journalistenfonds des Journalistenverbandes von Belarus, 1995), 202–27.

Sahm, Astrid, and Manfred Sapper and Volker Weichsel (eds). 'Tschernobyl: Vermächtnis und Verpflichtung', special issue of *Osteuropa*, 56(4) (2006).

Sandalls, F.J. *Hot Particles from Chernobyl: A Review.* N.p.: Great Britain, Ministry of Agriculture, Fisheries and Food, 1991.

Schmid, Sonja. *Producing Power: The Pre-Chernobyl History of the Soviet Nuclear Industry.* Cambridge, MA: MIT Press, 2015.

———. 'When Safe Enough Is Not Good Enough', *Bulletin of the Atomic Scientist* 67(2) (2011): 19–29.

SCPRI. *Organisation pratique de la radioprotection en France dans le cas de la santé publique et du travail. (Rapport SCPRI no 75).* N.p.: SCPRI, August 1962.

Simonnot, Philippe. *Les nucléocrates.* Grenoble: Presses universitaires de Grenoble, 1978.

Simons, Paul. 'Ill Winds Carried Chernobyl Poison', *The Times*, 26 April 2006.

Slovic, Paul. 'Perception of Risk', *Science* 236(1987): 280–85.

Smith, F.B., and M.J. Clark. *The Transport and Deposition of Airborne Debris from the Chernobyl Nuclear Power Plant Accident with Special Emphasis on the Consequences to the United Kingdom.* Bracknell: Meteorological Office, 1989.

Snape, Tony. 'Chernobyl Pay-Out Reaches £1.7 m', *The Observer*, 17 September 1989, p. B5.

Stead, Jean. 'Scottish Fallout Four Times Critical Levels', *The Guardian*, 9 May 1986, p. 2.

———. 'Radioactivity Alarms Scotch Makers', *The Guardian*, 7 May 1986, p. 2.

Stone, Philip R. 'Dark Tourism, Heterotopias and Post-Apocalyptic Places: The Case of Chernobyl', in Leanne White and Elspeth Frew (eds), *Dark Tourism and Place Identity* (Melbourne: Routledge, 2013), no page numbers for online edition.

Strazzulla, Jérome, and Jean-Claude Zerbib. *Tchernobyl – Les médias et l'événement.* Paris: La Documentation française, 1991.

Streiff, Gérard. 'Le syndrome de Tchernobyl', *L'Humanité*, 30 April 1986.

Stsiapanau, Andrei. 'The Chernobyl Politics in Belarus: Interplay of Discourse-Coalitions', *Filosofija. Sociologija* 21(2) (2010): 143–51.

Sussex Energy Group. 'Response to Government's "The Future of Nuclear Power" Consultation', 2007. Available online: https://www.sussex.ac.uk/webteam/gateway/file.php?name=seg-spru-nuclear-response.pdf&site=264 (last accessed 15 February 2020).

Tardieu, Vincent. 'L'état de l'hexagone après le nuage', *Libération*, 9 May 1986.

———. 'La France miraculée', *Libération*, 7 May 1986.

———. 'Catastrophe dans une centrale nucléaire soviétique', *Libération*, 29 April 1986.

Taylor, Richard Norton. 'Chernobyl Cloud Jams Whitehall Phones', *The Guardian*, 8 May 1986, p. 2.

Tchertkoff, Wladimir. *Le crime de Tchernobyl: Le goulag nucléaire.* Arles: Actes Sud, 2006.

Teräväinen, Tuula, and Markku Lehtonen and Mari Martiskainen. 'Climate Change, Energy Security, and Risk – Debating Nuclear New Build in Finland, France and the UK', *Energy Policy* 39(6) (2011): 3434–42.

TF1. 'Interview with Pierre Pellerin', *13 h*, 10 May 1986.

———. 'Interview with Pierre Pellerin', *13 h*, 29 April 1986.

The Chernobyl Forum. *Chernobyl's Legacy: Health, Environmental and Socio-Economic Impacts and Recommendations to the Governments of Belarus, the Russian Federation and*

Ukraine. Vienna: IAEA 2006 (Second Revised Version). Available online: https://www.iaea.org/sites/default/files/chernobyl.pdf (last accessed 15 February 2020).

The Daily Record. 'Nuke Link to Cancer Ruled Out', 23 April 1996.

———. 'Fags to Blame for Isle Cancer Scare - Health Chief's Claim', 2 April 1996.

The Greens. 'Windscale 1957...', *The Guardian*, 8 May 1986, p. 4.

The Guardian (online). 'Chernobyl Nuclear Disaster – in Pictures', 26 April 2011: http://www.theguardian.com/environment/gallery/2011/apr/26/chernobyl-nuclear-disaster-in-pictures (last accessed 15 February 2020).

The Guardian. 'Disaster that Fell with the Rain on a Bleak Hill', 27 April 1990, p. 25.

———. 'Chernobyl Bans Criticised', 19 September 1989, p. 2.

———. 'Baker Acts to Soothe Radioactivity Fears', 7 May 1986, p. 6.

———. 'A Disaster without Frontiers', 30 April 1986, p. 10.

———. 'PM Rejects Sizewell Rethink Call', 30 April 1986, p. 5.

The Independent. 'Experts to Investigate Island Cancer Deaths', 1 April 1996.

The International Chernobyl Project. *An Overview: Assessment of Radiological Consequences and Evaluation of Protective Measures. Report by an International Advisory Committee.* Vienna: IAEA, 1991.

The National Diet of Japan. *The Official Report of the Fukushima Nuclear Accident Independent Investigation Commission. Executive Summary,* 2012. Available online: http://www.nirs.org/fukushima/naiic_report.pdf (last accessed 15 February 2020).

The New York Times. 'Chernobyl Trials Begin', 8 July 1987, p. 5.

The Scotsman. 'MSPs Vote No to Nuclear Stations', 17 January 2008: https://www.scotsman.com/news/politics/msps-vote-no-to-new-nuclear-stations-1-1074436 (last accessed 15 February 2020).

The Sun. 'A Life Line to Radiation Kids', 25 April 2006.

The Telegraph (online). 'Only Two in Five Believe Climate Change Caused by Human Activity', 14 November 2009: https://www.telegraph.co.uk/news/earth/earthnews/6565955/Only-two-in-five-believe-climate-change-caused-by-human-activity.html (last accessed 15 February 2020).

Thorpe, Dave. *Doc Chaos: The Chernobyl Effect*. London: Hooligan Press, 1988.

Tinning, William. 'Chernobyl Fall-Out Denial over Benbecula Cancers', *The Herald Scotland*, 4 April 1996.

Tompkins, Andrew. *Better Active than Radioactive! Anti-Nuclear Protest in 1970s France and West Germany*. Oxford: Oxford University Press, 2016.

Topçu, Sezin. *La France nucléaire: L'art de gouverner une technologie contestée*. Paris: Seuil, 2013.

———. 'L'agir contestataire à l'épreuve de l'atome: Critique et gouvernement de la critique dans l'histoire de l'énergie nucléaire en France, 1968-2008'. PhD dissertation. Paris: Ecole des Hautes Etudes en Sciences Sociales, 2010.

———. 'Confronting Nuclear Risks: Counter-Expertise as Politics Within the French Nuclear Energy Debate', *Nature and Culture* 3(2) (2008): 225–45.

———. 'Les physiciens dans le mouvement anti-nucléaire: entre science, expertise et politique', *Cahiers d'histoire. Revue d'histoire critique* 102(2) (2007): 89–108.

Touraine, Alain et al. *La prophétie antinucléaire*. Paris: Seuil, 1980.

Travis, Alain. 'No Radiation Threat to the UK, Commons Told', *The Guardian*, 30 April 1986, p. 4.

Tucker, Anthony. 'Levels Dropping Fast as Cloud Moves away', *The Guardian*, 7 May 1986, p. 2.
———. 'Europe Fears Effects of High Radiation', *The Guardian*, 6 May 1986, p. 1.
———. 'Unquenchable Core of Fear inside the Poisoned Furnace', *The Guardian*, 30 April 1986, p. 23.
Turner, Royce Logan. 'Let the Campaign Roll Again', *The Guardian*, 12 May 1986, p. 9.
Twigge, Stephen and Len Scott. *Planning Armageddon: Britain, the United States and the Command of Western Nuclear Forces, 1945–1964*. Amsterdam: Harwood Academic Publishers, 2000.
UNDP and UNICEF. *The Human Consequences of the Chernobyl Nuclear Accident – A Strategy for Recovery*. New York: United Nations, 2002.
UNSCEAR. *Sources and Effects of Ionizing Radiation, UNSCEAR 2000 Report to the General Assembly with Scientific Annexes. Annex J: Exposures and Effects of the Chernobyl Accident*. New York: United Nations, 2000.
van der Pligt, Joop. *Nuclear Energy and the Public*. Oxford: Blackwell, 1992.
Vletch, Andrew. 'Parents Scour Chemists Shops for Iodine Tablets', *The Guardian*, 2 May 1986, p. 6.
Vodnik, Pierre. 'Le syndrome du nuage panique les Polonais', *Libération*, 2 May 1986.
Vulliamy, Edward, and Tony Heath. 'Chernobyl Controls on Sheep Extended', *The Guardian*, 13 August 1987, p. 1.
Wainwright, Martin. 'Ministers Deny Risk as Fallout from Chernobyl Lingers in Sheep', *The Guardian*, 30 December 1994, p. 9.
Wales on Sunday. 'Remembering Chernobyl', 23 April 2006.
Walker, Linda. 'Die Arbeit von "Chernobyl Children's Project (UK)" und ein Überblick über weitere Tschernobyl-Organisationen und Netzwerke in Großbritannien', in IBB (ed.), *Tschernobyl und die europäische Solidaritätsbewegung* (Norderstedt: Books on Demand, 2011), 108–16.
———. *Living with Chernobyl: Ira's Story*. Tunbridge Wells: Ticktock Media, 2005.
Walker, Martin. 'Silence Covers "Zone of Death"', *The Guardian*, 30 April 1986, p. 1.
———. 'The Vision Consumed in the Fire this Time', *The Guardian*, 30 April 1986, p. 23.
Walker, Martin, and Edward Valliamy. 'Airport Radiation Tests for Students', *The Guardian*, 2 May 1986, p. 1.
Ward, David. 'Hill Farmers Living under Cloud as Effects of Chernobyl Linger', *The Guardian*, 25 April 1994, p. 7.
Warren, Jane. *Igor: The Courage of Chernobyl's Child*. London: Boxtree, 1997.
Watt Committee on Energy. *Five Years after Chernobyl. 1986–1991: A Review*. London: Watt Committee on Energy, 1991.
Wavell, Stuart, and Stephen Cook. 'Case of Missing Isotope', *The Guardian*, 7 May 1986, p. 32.
Weart, Spencer R. *Nuclear Fear: A History of Images*. Cambridge, MA: Harvard University Press, 1988.
Welsh, Ian. *Mobilising Modernity: The Nuclear Moment*. London: Routledge, 2000.
———. 'The NIMBY Syndrome: Its Significance in the History of the Nuclear Debate in Britain', *British Journal for the History of Science* 26(1) (1993): 15–32.
Wendland, Anna Veronika. 'Reaktorsicherheit als Zukunftskommunikation: Nuklearpolitik, Atomdebatten und kerntechnische Entwicklungen in Westdeutschland und Osteuropa

1970–2015', *in* Christoph Kampmann et al. (eds), *'Security Turns Its Eye Exclusively to the Future': Zum Verhältnis von Sicherheit und Zukunft in der Geschichte* (Baden-Baden: Nomos, 2018), 305–35.

———. 'Rezension zu: Kalmbach, Karena: Tschernobyl und Frankreich. Die Debatte um die Auswirkungen des Reaktorunfalls im Kontext der französischen Atompolitik und Elitenkultur. Frankfurt am Main 2011', H-Soz-u-Kult, 20 August 2012: http://hsozkult.geschichte.hu-berlin.de/rezensionen/2012-3-086 (last accessed 15 February 2020).

Western Daily Press. 'Nature's Cure for Nuclear Fallout: Wildlife Heaven', 21 April 2006.

White, Michael, and Martin Walker and Alex Brummer. 'US Estimates up to 3,000 Victims from Satellite Information', *The Guardian*, 1 May 1986, p. 1.

WHO, *Health Effects of the Chernobyl Accident and Special Health Care Programmes, Report of the UN Chernobyl Forum Expert Group 'Health'*. Geneva: WHO, 2006. Available online: https://www.who.int/ionizing_radiation/chernobyl/who_chernobyl_report_2006.pdf (last accessed 15 February 2020).

Wikipedia. 'Ali Hewson': http://en.wikipedia.org/w/index.php?title=Ali_Hewson&oldid=480318803 (last accessed 15 February 2020).

———. 'Independent WHO': http://fr.wikipedia.org/wiki/Independent_WHO (last accessed 15 February 2020).

———. 'Talk: Christopher Busby': http://en.wikipedia.org/wiki/Talk:Christopher_Busby#cite_note-0 (last accessed 15 February 2020).

World Nuclear Association. *Fukushima and Chernobyl: Myth versus Reality*. London: WNA, 2012. Available online: https://www.world-nuclear.org/focus/fukushima/fukushima-and-chernobyl-myth-versus-reality.aspx (last accessed 15 February 2020).

Worley, Norman, and Jeffery Lewins (eds). *The Chernobyl Accident and Its Implications for the United Kingdom*. London/New York: Elsevier Applied Science Publishers, 1988.

Wynne, Brian. *Rationality and Ritual. Participation and Exclusion in Nuclear Decision-Making*. Abingdon: Earthscan, 2011.

———. 'Misunderstood Misunderstandings: Social Identities and Public Uptake of Science', in Alan Irwin and Brian Wynne (eds), *Misunderstanding Science? The Public Reconstruction of Science and Technology* (Cambridge: Cambridge University Press, 1996): 19–46.

———. 'Misunderstood Misunderstanding: Social Identities and Public Uptake of Science', *Public Understanding of Science* 1(3) (1992): 281–304.

———. 'Sheepfarming after Chernobyl. A Case Study in Communicating Scientific Information', *Environment* 31(2) (1989): 10–39.

Yablokov, Alexey, and Vassily B. Nesterenko and Alexey V. Nesterenko. *Chernobyl: Consequences of the Catastrophe for People and the Environment*. Boston: Blackwell Publishers on behalf of the New York Academy of Sciences, 2009.

Yaroshinskaya, Alla. *Chernobyl: The Forbidden Truth*. Oxford: Jon Carpenter, 1994.

———. *Tchernobyl, vérité interdite*. La Tour-d'Aigues: Editions de l'Aube, 1993.

Young, Hugo. 'Crisis of Credibility Exposed by Chernobyl', *The Guardian*, 8 May 1986, p. 21.

Zink, Andrea. 'Approaching the Void: Chernobyl in Text and Image', *Anthropology of East Europe Review* 30(1) (2012): 100–12.

Index

www.ingramcontent.com/pod-product-compliance
Ingram Content Group UK Ltd.
Pitfield, Milton Keynes, MK11 3LW, UK
UKHW021837100926
8556IPUK00004B/406

9 781789 20702